This book is also dedicated to my father William Zhuang,

who encouraged me to strive and make a difference to this world.

建筑工程口语实战攻略

Practical Strategies for Oral Engineering English

庄从容◎编　著

浙江大学出版社

图书在版编目(CIP)数据

建筑工程口语实战攻略 / 庄从容编著. —杭州：浙江大学出版社，2019.11

ISBN 978-7-308-19360-3

Ⅰ. ①建… Ⅱ. ①庄… Ⅲ. ①建筑工程—英语—口语

Ⅳ. ①TU

中国版本图书馆 CIP 数据核字(2019)第147804号

建筑工程口语实战攻略

庄从容 编著

责任编辑 李 晨

责任校对 郑成业

封面设计 项梦怡

出版发行 浙江大学出版社

(杭州市天目山路148号 邮政编码310007)

(网址：http://www.zjupress.com)

排 版 杭州朝曦图文设计有限公司

印 刷 嘉兴华源印刷厂

开 本 787mm×1092mm 1/16

印 张 10.75

字 数 330千

版 印 次 2019年11月第1版 2019年11月第1次印刷

书 号 ISBN 978-7-308-19360-3

定 价 36.00元

序 言

在多年的英语教学生涯中，我一直在问自己这样一个问题：一名来自企业的学员究竟需要什么？在我转入企业培训领域最初的三年中，我更加深入地思索着这个问题。因为每当回访来自企业的学员时，我总会被几乎相同的回答所“伤害”，学员们常说：“老师，实在对不起，我把之前学的东西又还给你了！”“因为平时工作完全用不上。”这难道是长时间培训后该有的结果吗（通常企业培训会持续若干周或若干月）？我能够理解，大部分企业的员工在学习时间和工作时间上难以平衡，加上生活当中能用到这些语言内容的地方少之又少，这令他们无法对学到的语言知识保持敏感。究其症结，问题根源也许就在于课程与教材本身。

在传统的企业培训过程中，无论面对哪个年龄阶段、哪种工作背景的学员，我们所使用的教材都是讲授以日常生活为主的英语知识。大部分从事英语教学的老师在课程当中都只围绕日常英语进行教学，特别是在口语教学的过程中，无论是课堂上的内容、活动组织，还是场景演练，都围绕着生活话题和学习话题展开，以培养兴趣为主。因为这样的课程有普及性，对于老师来讲也非常容易上手。并不是说还有什么问题，但我认为，英语语言的学习，特别是职场人的语言学习，应当侧重于语言的实用性和专用性。课程及教材能让学员们在培训后立刻将其所学使用到日常的工作中。

因此，在最近几年的企业培训中，我尝试着在巧用传统口语课程精华内容的基础上，结合不同公司的行业需求，将相应的专业词汇融入教学内容，将各个不同的专业场景融入课堂口语练习，让课程效果发生了质的改变。因为学员们在学习和课堂训练的过程中，都会联想到他们日常的工作，并且再次确认工作中的英文用法。

我以培训师身份加入中电四公司（中国电子系统工程第四建设有限公司）的日子，也是我将传统教材打碎重建、让自己教授的课程重生的过程。在长期的培训教学过程中，我针对工程师们的日常工作和专业需求进行了系统性的深入了解与研究，着力搜集编辑通用性和实用性较强的专业词汇、场景对话，根据工程师群体的学习特性设计课程方案，在教学实践中不断完善教材的编撰。

这是一本专门为工程企业量身定制的英语培训课程配套的标准化教材。

本书围绕着工程师们日常最基本的英语交流，话题与场景包括从提出疑问，到PPT演讲技巧、项目执行、工程采购、解决纠纷等，融入核心的口语与语法内容，让学员们学会在专业场合中更准确地用英文表达自己。

与其他一些专业教材不同的是，本书并非将专业词汇知识生搬硬套地融入教学当中。每节课的课堂内容、课后活动都能让学员在全过程中保持浓厚的学习兴趣和高昂的参与活力，在快速提高学员英语运用能力的同时，促进了学员今后英语学习和运用的习惯养成，收到一次培训、受益终身的效果。

随着中国经济的快速发展，越来越多的企业也愈发外向，建筑类公司不但在向外拓展，也面临着与国际公司在本土的竞争，这无疑将我们放入了一个世界级的“竞技场”，而英语作为国际通用的主要语言工具，则是我们首先需要拿到的敲门砖和重要武器。我们要努力培养出不但在专业上与国际接轨，同时在语言上也能与国际接轨的人才来为我们的行业提供更好的服务。这就要求我们的教学内容能够更加有的放矢，更加注重学员的专业需求，让我们更有效率地完成人才语言培养的任务。这是一项崇高的历史使命，本书的出版意在为此项历史使命添砖加瓦。

特别感谢James Anthony Thorpe先生和庄艳玲老师，以及所有为本书提供帮助与智慧的中国电子系统工程第四建设有限公司的同仁们，是你们让这本书成为可能！

祈阅读本书的专家学者和使用本书的教师以及学员们不吝指教。感谢您的阅读！

庄从容

2019年3月1日

Prologue

Through years of English teaching, I kept asking myself the same question: "What kind of English training does a corporal employee need?" In the first three years I worked as a trainer specializing in corporal training, I pondered deeper into the question because every time when I was doing an after-training survey, I was always "harmed" by the same reply over and over again. "Sorry Richard, I have returned you everything you taught me! I just cannot use them during my actual work!" Is this the result of our long-term training?(Normally it would take weeks or months to complete a corporal training.) Just think about the time we might have spent on the training! I can understand the fact that it is extremely difficult for most corporal employees to divide their time of work and language study evenly, and the lack of language environments makes it difficult for them to insist on language learning, and I believe that the biggest source of the problem comes from what is taught in class and the direction of our learning materials.

Traditionally, English corporal training's class content and materials are normally focused on daily English regardless of the student's needs and their professional backgrounds. There is nothing wrong with it, because the basic part of training class is daily English, which is quite easy to organize, with role playing and building up student's interest. Such training can be said both universal and simple to grasp. However, it is to my understanding that language learning, especially for professional personnel, should be more practical with an emphasis on their profession, which would be instantly applicable to student's daily work after the training.

Therefore, in recent years I intentionally combine the professional demands of different corporations together with the essence of traditional oral English class contents to create a "fusion" class that has all kinds of different professional scenarios. That essentially changed the value and impact of the class. Because in this way, students can always relate language learning to their daily work during the training process and reaffirm the English terms through the training.

It was during my years as a trainer in CEFOC (The Fourth Construction Co., Ltd. of China Electronics System Engineering), one of the most influential EPC (Engineering Procurement Construction) firms in China, that I started the process of breaking and reassembling traditional teaching methods into a brand new style. Through the teaching years of training, I made deep analysis into the actual professional language demands of our engineers, and collected "most-use" words and expressions as well as scenarios to design classes and learning materials for engineers, while continuously perfecting the edition of this book.

This English training textbook is a standardized textbook as a result tailored for engineering corporations.

The book focuses on the most fundamental daily English scenarios of engineers, from asking questions, to power point presentation, project execution, procurement, problem solving and all kinds of engineering scenarios in combination with oral English and grammar contents to enable our students to express themselves more accurately.

Different from some other professional textbooks, I did not squeeze all the professional vocabularies and knowledge straight into the content of the book. Through activities in each class, the training course can constantly maintain the interest and vitality of the students, which greatly increases our students' practical English ability as well as their learning habits. The purpose of this book is to create a class that you only need to attend once, and can benefit from it for your entire career.

It is undeniable that China's economy has been developing rapidly nowadays, which means more and more corporations are willing to expand their business overseas. As for construction firms, not only would they try to expand their business internationally, but they would also have to face the competitions of international companies domestically, which will put them into the international "arena" of business. As a universal language, English would be the most critical weapon for us to struggle in the international market. Our effort is to train professional employees well adapted to the international market as well as language expertise who can connect to international market and provide better services for the industry. Therefore, our class content needs to be more precise, and emphasize on our students' professional demands, because this would complete the task of training talents with language expertise much effectively. I believe this book is an effort to the great mission. I truly wish that through the publication of this book, we could give more contribution to this mission.

Special thanks to Mr. James Anthony Thorpe, Mrs. Zhuang Yanling, and all the colleagues from CEFOC who provide support and wisdom for this book, and you guys make it possible!

Moreover, I sincerely look forward for more guidance from professionals and experts if possible! Thank you for reading!

Richard Zhuang

March 1, 2019

本书使用说明

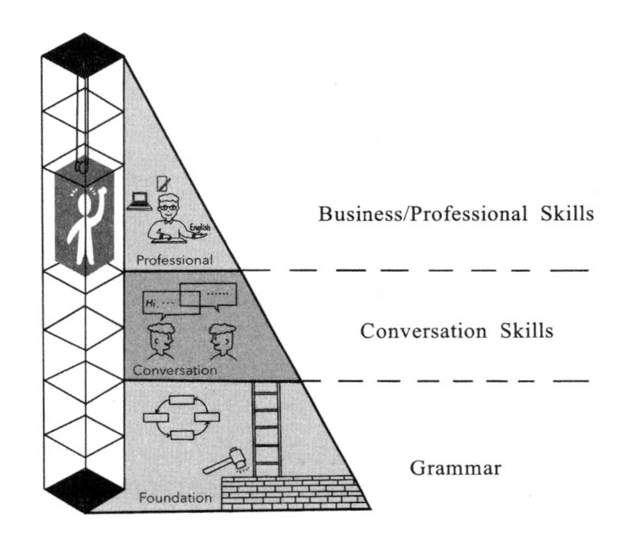

本书的学习逻辑可以用上方金字塔的形式来呈现。首先，我认为看似杂乱的沟通过程其实都围绕着一个中心的地基，那就是语法。每种不同的语法都携带着不同的目的，比如现在时陈述事实，一般过去时陈述的是具体的时间发生过的事件，将来时多数代表了计划和目标。灵活使用它们能够让你在沟通的过程中可以更为准确地表达自己的观点。那么学习本书的第一步一定是从口语语法入手，学会每一课中最常用的语法结构是完成课程要求最为核心的部分。每节课中的核心知识点(core knowledge)基本都是重要的语法点，领会它能让你学得更有效。

金字塔打好了地基，那么就可以往上爬了。下一步就是对话技巧。什么时候，什么场景，针对什么情况应说些什么则是我们的第二个台阶。每一节课都有一个大主题和三个小主题，每一个主题都有一个相应的场景。各种不同的场景都以对话的形式来呈现每个知识点在实践中的正确运用方式。这是一个从书面到实践的台阶，是语法与运用融合的台阶。

金字塔的最高层是商务/专业技巧，就是我们将自身专业知识、专业词汇与场景、语法知识点进行一个大的融合。这是金字塔的尖端，这一步要求学员能够将日常的工作与学

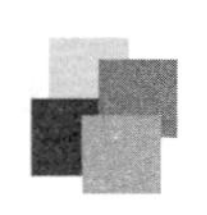

到的知识进行专业化、个人化融合。本书的每一课都有工程专业的场景,每课结尾也有相应的专业知识附录。在这一步,学员必须进行大量的角色扮演,每课的角色扮演活动的答案都是开放的,学员们应当根据个人的专业来进行回答,创造尽可能多的答案。这样学员才能更灵活、更专业地运用所学的知识。

因为沟通没有唯一的答案。

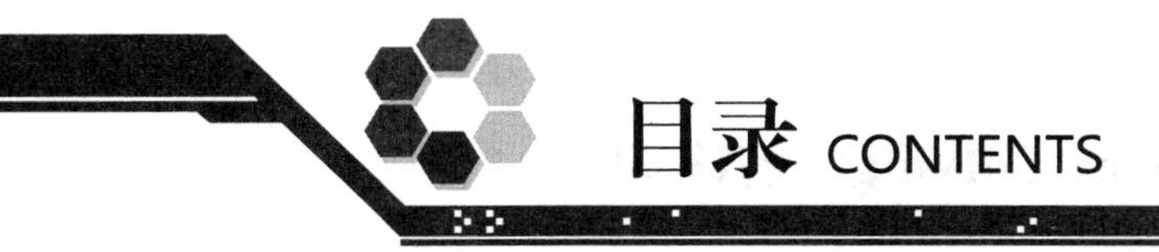

目录 CONTENTS

Class 1 First Day on the Project 项目第一天

Class 2 Experience 谈谈经验

Class 3 Opinions to the Case 看法

Class 4 Engineering Presentations 工程类演讲

Class 5 Problem Solving 解决问题

Class 6 Business Changes 行业变化

Class 7 What Would You Do? 你会怎么做?

Class 8 Socializing 社交

Class 9 Procurement 采购

Class 10 Promises and Regulations 承诺与规章制度

Class 11 Meeting Minutes 会议纪要

Class 1

First Day on the Project

项目第一天

Chapter A Greetings 大家好

 Ⅰ Key Words and Expressions

company 公司	How is work? 工作如何?	chief-engineer 总工程师	quality inspection 质量检查
business 商务,生意	Good morning/afternoon/evening. 早安/中午好/晚安。	Introduce yourself/ourselves 介绍自己/大家	have been doing 一直在做
intern 实习生	EHS (environment, health, safety) 环境、健康、安全	fantastic/great/pretty good 赞/棒/挺好的	How are you doing? 过得怎样?
How is everything going? 一切怎样?	EHS engineer 安全工程师	Follow me. 跟我来。	How is it going? 最近怎么样?
project site 项目现场	Anything up? 有什么新鲜事吗?	get used to 习惯	Write down any other words you don't know and ask.

 Ⅱ Read the Material Below and Answer the Questions

Peter: Morning guys, welcome to the PPG project site. I'm Peter, your guide for today, and I'm sure it's the first time you guys are here. Why don't we say hi to each other and introduce ourselves first? Yes, what's your name? How are you, sir?

Jason: I'm Jason Lee, from China... Sorry, my English is not so good.

Peter: Well, I'm sure you could tell us a bit more than that. And you, sir, how are you doing?

Ken: Pretty good! I'm Ken Stevenson. I'm an EHS manager. Nice to meet you!

Peter: Well, hello Ken, what about this young lady here? How is everything going?

Maria: Fantastic! I'm Maria. This is my first time here, and I'm so excited! I'm an EHS engineer, and I have been doing this work for the past 2 years. Also I have worked for the PPG company since I was an intern!

Peter: Great! That's quite a nice introduction you give us here! I'm sure you could get used to here pretty quickly. Next, I'm going to show you our quality inspection office. Please follow me.

Questions

1. What are the speakers doing on the site?
2. Who do you think speaks English best here? Why?

 Ⅲ Core Knowledge for Greetings

当你与陌生人说话的时候,你第一句话会说什么?当陌生人与你说话的时候,他的第一句话又会是什么?答案非常明确,那就是打招呼。通常我们打招呼的时候都会用hello和how are you,而对其他的招呼用语十分生疏,这给我们与外国人交流带来了很大的困扰,因为英语中打招呼的方式多种多样,那么我们该如何来回应这些招呼呢?

我们可以把招呼分成两个不同的组别。

问题式招呼	回应
How are you?	Good.
What's going on?	Fantastic.
How's it going?	Great.
How are you doing?	Pretty nice.
How's everything going?	Not bad.
What's up?	Wonderful.

我们可以发现,问题式的招呼都是以问句起始,以how和what作为开头,而且需要一个答案。而答案式的招呼,则仅仅用同样的话回应即可。

答案式招呼	回应
Good morning.	Morning.
Good afternoon.	Afternoon.
Good evening.	Evening.
Hello.	Hello.
Hi.	Hi.

Tip:如果不想记那么多的回应方式,那么只需要拿其中的一个作为你的专属用语就行了! E.g. A: How are you doing? B:Good! 同时,还可以加上You?(你呢?)来反问对方,延展对话。

 Ⅳ Exercise and Activity

Practice these greetings below with a smile on your face!

E.g. How are you doing?

I'm pretty good, you?

Morning!	Hey!	What's going on?
Afternoon!	Good day!	Anything new?
Evening!	Anything up?	How's the project?
How's going?	How is your stay here?	What's up?
How are you doing?	How is your work?	Anything up lately?

Chapter B Learning to Ask Questions 问问题

I Key Words and Expressions

SIEMENS 西门子	expensive 昂贵的	easier 更简单的	complete 完成	had a raise 加薪	cost 开支
owner 业主	labor 劳工	task 任务	extremely 极度地	investment 注资	major 主要的,专业
area 地区,区域	cheaper 更廉价的	hire 雇佣	land cost 地价	schedule 日程	suggestion 提议
located 位于	demanding 苛刻的	recently 最近	Catch up with you later. 一会见。	high standard 高标准	local 本土的,当地的
bid 竞标	not easy to get along with 不好相处	Good luck on your work. 祝你工作顺利。	useful information 有效信息	Best city on earth! 世上最好的地方!	Write down any other words you don't know and ask.

II Read the Material Below and Answer the Questions

Lawson: So, Pat, how is work recently?

Patrick: Not bad, we have just won the bid for SIEMENS.

Lawson: Great! So where is the project site located?

Patrick: It's going to be built in the Changping area in Beijing.

Lawson: Why Changping?

Patrick: Because the land cost is not really expensive, and it's easier for us to hire cheaper labors around the area.

Lawson: So, who is the owner? Is he a nice guy?

Patrick: Mr. Hans Morgan is the owner. He is not easy to get along with, an extremely demanding boss. He gave us huge tasks with so little time to complete.

Lawson: Well, I guess I have to go now. Really nice talking to you. Good luck on your work!

Patrick: You too. Catch up with you later.

Questions

1. What are they talking about?
2. Why did the owner choose Changping area for the project?
3. Did Lawson give Patrick any useful information?

 Ⅲ Core Knowledge for Asking Questions

1. How to answer them(如何回答)?

Open Questions		Close Questions (Speculation)	
What...?	It's... I think it's...	Do you...?	Yes, I do. / No, I don't.
Why...?	Because...	Are you...?	Well, I am. / I am not.
(1) How to (解决方式)...? (2) How much...? (3) How are...?	(1) We could... We can... (2) It's 5 dollars. (3) Pretty good.	Normally just "yes" or "no".	
Who...?	I'm... He is...		

2. How to make a chain of questions in a conversation (如何建立对话问题链)?

Work	Interest	Education	Family
How is your work?	How is your stay in Beijing?	Where did you go to school?	Where are you from?
What do you think about our company?	Any plans after work?	What's your major?	Can you tell me about your hometown?
Do you have any suggestions for us?	Do you enjoy any local dishes?	What's your major about?	How is your family?

我们将各种各样的话题分为四组,然后在对话过程中选择一组作为起始话题,比如说我选择工作作为话题。

A: So, how is your work?

B: Well, the project was going quite well, and I had a raise last week.

A: Great! So what do you think about your project?

B: Well, the project CORNING is quite a mature and high standard project, and the investment was just large.

A: Any suggestions for our project in Changping?

B: Well, I guess you guys need more labor on the site. It's a bit over the schedule.

问完三个关于工作的问题后,话锋一转,可以选择下一个话题,比如interest (兴趣)。

A: So, how is your stay in Beijing?

B: Absolutely wonderful. Beijing is the best city on the earth!

A: Any plans after work today?

B: Nope. Let's go somewhere for fun!

如果对方还兴趣盎然,那么可以往下继续问关于家庭,或者教育方面的更深入的问题。Tip:当我们使用开放式问题(6W1H)的时候,会发现虽然问题是一样的,但是答案都是不同的。

3. How to end a conversation(如何结束对话)?

Every conversation needs an ending, so how do we end a conversation before you are speechless? Here are some expressions that you can use.

In a Restaurant	After Work Conversation	Others
Cheers, for good business.	Well, let's call it the day! Catch you up later.	Have a good stay in China! Well, have a good day!

 Ⅳ Exercise and Activity

回答下列问题,并与小组成员互相提问,不要让对话停下来。

How is your work?	
What do you think about your company?	
Do you have any suggestions for us?	
Where did you go to school?	
What's your major?	
Where are you from?	
Tell me about your hometown.	
How is your family?	
How is your stay in the city?	
Any plans after work?	

Tip: Use "So, Well, OK" to start your conversation. (So, Well 和 OK 在此表示"那个……",这会让听众注意你接下来要说的话。)

 Chapter C Working Status and Plans 工作状态和计划通知

 I Key Words and Expressions

usually 平常	pump myself up 给自己补充能量	hand something to someone 交给
close a deal 签订协议	working experiences 工作经验	catch up with 赶工
cost engineer 成本工程师	disagree 不同意	construction phase 工程阶段
stressed out 累坏了	documents 文件	prediction 预测
energetic 精力充沛	organize 整理	inspection/investigation 检测/调查
drawings 图纸	eventually 最终	Write down any other words you don't know and ask.

Ⅱ Read the Material Below and Answer the Questions

Joyce: Well, I usually get to work at 8:00 a.m., and I never get to work late. Right now, we are closing a deal on a big case. So as a cost engineer I have to work till 9:00 p.m. to finish. The work is getting many of us tired and stressed out, but I'm still very energetic.

Dwight: Well, now I'm trying my best to catch up with my work because I need to finish my drawings and documents before deadline, and I'm drinking a lot of energy drinks today to pump myself up. For now, I need some sleep!

Mark: I'm the CEO here. I work 6 days a week, and I have meetings every day. I often work overtime and so do my employees. Recently, one client asked us to change our design and disputed our contract, so I think I'm going to cancel my Spring Festival holiday to fix it.

Charlie: Well, I'm the son of Mark. So I think that I could do anything that I want, and I'm not doing any actual works. I'm just staying here because I need some "working experiences" to make my dad happy. Eventually he is going to hand this company to me.

Questions

1. What's Mark's plan for the Spring Festival? Why?
2. Who is the best employee here? Why?

Ⅲ Core Knowledge About Simple Present, Present Continuous, and Future Tense

1. 三种时态的表意功能，如下。

Simple Present	Present Continuous	Future
current condition	temporary events	plans
repeated events	on-going events	solution
routines	now events	prediction
truth		promise

2. 状语对于语法不好的人来说几乎是“救命稻草”。口语表达时，在时态使用错误的情况下，可在时态后用状语来修饰时间，让对方明白你具体想表达的时间和状态，同时也可以用于强调时间。

Simple Present	Present Continuous	Future
usually	now	tomorrow
normally	today	next week
everyday	this week	in year 2025

3. 三种时态在语法中的表现，如下。

Simple Present	am/is/are/verb（动词）
Present Continuous	am/is/are+*v*.-ing
Future	will do, am/is/are going to do

Tip: Before we talk about something, we must first know the goal of this conversation, because the tense not only represents a time, but also a goal.

Ⅳ Exercise and Activity

1. Describe these pictures below using simple present, present continuous, and simple past.

Who is he?

What does he do? What's his job?

What is he doing now?

What is he going to do?

2. Make plans for a one day construction training.

8:00-8:30 a.m.	We are going to have a morning meeting to take a look at the project.
8:30-10:00 a.m.	
10:00-12:00 a.m.	
1:30-3:00 p.m.	
3:00-4:30 p.m.	
4:30-9:00 p.m.	

（趁热打铁，课后练习见143页）

Appendix EPC Task Sheet 总承包作业清单

1. This is an EPC plan for a pharmaceutical project.

Task	Start	Finish	Duration	Actual Time
planned EPC EPC 计划	January 1st	February 2nd	33 days	40 days (Feb. 9th)
design 设计				
authorization 授权				
detailed engineer 细节设计				
procurement 采购				
construction 建设				
start up 启动				
start production 开始生产				
shut down 关闭				

2. Think about the possibilities that caused the delay.

A: Why was the design phase delayed for 2 weeks?

B: The delay was caused by...

(Write down more possibilities for the delay.)

...

Tip: 每当遇到麻烦或者争端的时候，不要先道歉，要先调查(investigate)。根据调查结果再决定责任归属和应对的措施。所以 The problem was caused by...或者 The problem happened because...都是比较常用的表达方法。

Answers

Chapter A Greetings

Ⅱ Read the Material Below and Answer the Questions

1. They were visiting the PPG site, and introducing themselves.

2. Obviously Maria and Peter. Peter talked a lot about the project, and Maria talked a lot about herself.

Chapter B Learning to Ask Questions

Ⅱ Read the Material Below and Answer the Questions

1. They are talking about the Changping project.

2. Because the Changping area has cheaper labors and lower land price.

3. No, he didn't. He is just asking many open questions.

Ⅳ Exercise and Activity

How is your work?	Well, not bad. Quite busy recently.
What do you think about your company?	Well, I think my company is quite mature and experienced in the engineering field.
Do you have any suggestions for us?	Not really, but it would be a good idea to make a KPI list for your employees.
Where did you go to school?	I graduated from the Cornell University in the USA.
What's your major?	My major is HVAC (暖通) system.
Where are you from?	Well, I'm from a city called Brisbane in Australia.
Tell me about your hometown.	Actually it's the most interesting place on the earth. You could find all kinds of animals there.
How is your family?	Everything has been quite good recently; my sister just had a baby last week.
How is your stay in the city?	Wonderful, I have tried some local dishes recently.

Chapter C Working Status and Plans

Ⅱ Read the Material Below and Answer the Questions

1. Mark is going to cancel the Spring Festival holiday because his clients want them to redo their design.

2. Joyce is the best employee, because she is energetic and organized.

Ⅳ Exercise and Activity

1. He is Jason.
   He is the designer of the FMT company.
   He is designing the pipeline of the KND project.
   He is going to send his design to the client tonight.
2. He is Danny.
   He is our crane operator (吊车司机).
   He is lifting a huge equipment.
   He is going to transport the euipment back to the building.
3. His name is Bowen.
   He is the chief engineer of our company.
   Now he is calling his collegues ro duble check the design.

Class 2

Experience
谈谈经验

Chapter A Report 报告

Ⅰ Key Words and Expressions

technical school 职业学院	teach me a lot 教会了我很多	newly engaged 新来的	project director 项目总监
Would you mind telling me...? 能否告诉我……?	northern branch 北方分公司	welder 焊工	doing a great job 干得很棒
tutor 导师	transferred 调任	employees 员工	truck drivers 卡车司机
technician 技工	interview 面试	negotiate 谈判	permit 许可
be awarded 荣获	installer 安装工	launch 启动	promotion 升职
system 系统	piper 管道工	chairman 主席,董事长	discuss the contract 讨论合同

Ⅱ Read the Material Below and Answer the Questions

Jamie: Good morning young man, you are new here.

Terry: Yes sir, I just graduated from the technical school.

Jamie: Would you mind telling me what you did for the last few days?

Terry: Actually, what I did mostly is learning how to operate the electrical systems of the project.

Jamie: Who is your tutor here?

Terry: The chief engineer Mr. Anderson is my tutor for this month. He taught me a lot about the electrical installation on the site.

Jamie: Old Anderson, good guy. He has excellent experiences in the field. So who is that girl over there? She also seems new here.

Terry: Her name is Lena. She is a technician. She has already been working in this industry for the past eight years.

Jamie: Oh well, it looks that she is doing a great job.

Terry: Sure she is. It was just last year, she was awarded the best engineer of the year because of her outstanding works, and she transferred to the northern branch this May.

Jamie: Wonderful. What about those young kids over there? I just saw them last Tuesday on a site interview.

Terry: Oh, they are our newly engaged pipers, truck drivers, installers, and welders.

Jamie: Seems like you know quite a lot about this project! Ah! Our project director is coming this way, let's go and say hi to him.

Questions

1. What did Jamie and Terry talk about?
2. Who is Lena? Why is she an excellent employee?

Ⅲ Core Knowledge of the Past Tense

1. 一般现在时与一般过去时的对比,如下。

时态	状态	行为	状语	表意功能
一般现在时	am/is/are	do	usually, normally	routines, truth
一般过去时	was/were	did	yesterday, last night, last year	specific events, report

2. 一般过去时的例句,如下。

was, were(过去的状态)	did(过去的行为)	was+v.-ing(过去进行中的行为)
She was the best employee last year.	She did some really good works last week!	She was working in my project office last Saturday.

Tip: Use simple past for information that is specific in the past or reporting events that were important back in the past.

3. 动词过去时变形,如下。

原形	规则过去式	原形	非规则过去式
construct	constructed	build	built
fix	fixed	give	gave
design	designed	speak	spoke
delay	delayed	become	became

4. 现在进行时和过去进行时的对比,如下。

时态	状语	表意功能
现在进行时	now, recently, today	temporary events
过去进行时	yesterday, last year	background action of the past

Ⅳ Exercise and Activity

1. 将以下动词转换为过去式并翻译。

launch		negotiate	
keep		solve	
cooperate		attend	
manage		arrange	
lose		organize	
win		engage	
control		permit	
spend		promote	
talk		document	
communicate		deal	

2. 用一般过去时回答以下问题。

Questions	Answers
Where was your last project located?	My last project was located in Shanghai.
Where did you work last week?	
Who was your first boss? Did you like him?	
When did you get off work yesterday?	
What major did you choose in school?	
Who was your favorite project manager?	

3. 阅读下文,回答以下问题。

On the 21st of September, we finished our engineering training, and on the next day the 22nd, our chairman visited us and gave us an unforgettable speech about his life. In the afternoon, our project manager had a meeting with our clients from the USA, and they discussed the contract with us. At night, our project manager invited them to have Chinese hot pot (火锅), and they surely enjoyed it.

What did we do on/at:

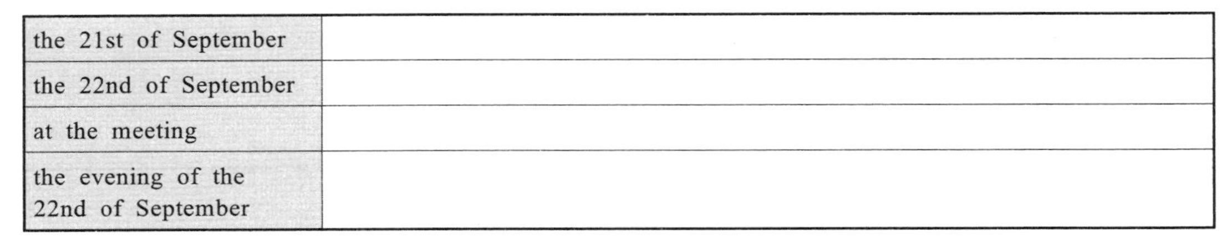

the 21st of September	
the 22nd of September	
at the meeting	
the evening of the 22nd of September	

Chapter B Experience and Achievements 经验与成就

Ⅰ Key Words and Expressions

go ahead 开始吧	achievements 成就	terrific 很棒	production center 生产中心
My pleasure. 我的荣幸。	absolutely 绝对地	photocopies 复印件	consult 顾问,咨询
representative 代表	focus on 专注于	qualification 资格证	main services 主要服务
brief introduction 简介	specialized 特别专业	brilliant 太棒了	hand over 拿过来,交工
start small 从小公司做起	research tower 科研塔	has what it takes to do something 拥有实力来做某事	general contractor 总承包人
revenue 利润	sound 听起来	satisfied 感到满意	sub-contractor 分包人

Ⅱ Read the Material Below and Answer the Questions

Jeffery: Hello ladies and gentlemen, I'm Jeffrey Houston, the representative today for CCOC, and it's a pleasure for me to give you a brief introduction about our company.

Paul: Well, go ahead, we are listening.

Jeffery: Well the first thing I'm going to talk about is our company's history. We have been working in the engineering industry for the past 20 years. We started small from 1998 and reached 200 million dollars worth of revenue last year.

Paul: Tell me about your services, and I want to know more about CCOC's experience and achievements.

Jeffery: Absolutely! EPC and EPCM are our primary services, and we have been focusing on and specialized in the electronic and pharmaceutical field from 2003. We have worked for famous companies such as the SIEMENS, CORNING, and Microsoft, and we have done projects such as the building of the SIEMENS Shanghai factory, the R&D building for BOE, and the research tower for the IVO Company.

Paul: Sounds terrific, and could you please hand over your qualifications for us to look at?

Jeffery: Sure, here are the photocopies of our qualifications. We have obtained qualifications from class-A general building contractor to Class-A M&E general contractor's qualification. You could have a look.

Paul: Brilliant, and Jeffery, what makes you think that CCOC has what it takes to provide excellent services to our project?

Jeffery: Excellency has always been the core of our practice ever since the first day we founded our company.

Questions

1. What projects has CCOC done?
2. Do you think Paul is satisfied with Jeffery's introduction? Why?

Ⅲ Core Knowledge of Talking About Your Experience and Achievements

1. 现在完成时和现在完成进行时的对比,如下。

时态	状语	表意功能
现在完成时	in the past/twice/for...years	actions in the past, experiences
现在完成进行时	since	the continuous action since the past to now, which still goes on

2. 现在完成时和现在完成进行时的"组合"用法,如下。

从过去到现在都在进行中的事业	在这之中所获得的成果和经验
Our company has been focusing on the pharmaceutical industry since 1991.	We have done pharmaceutical projects such as the novo nordisk and Roche production centers.
组合	
Our company has been focusing on the pharmaceutical industry since 1991, and has done pharmaceutical projects such as the novo nordisk and Roche production centers.	

这样不但介绍了公司一直以来坚持的事业,同时也介绍了公司的成果。

Tip: 有趣的是, 西方人在日常谈论自己的过去的时候最常用的时态是现在完成时,因为现在完成时所强调的是经历本身,而具体的时间或者状态是次要的。所以只要你做过,大可以用have done 来告诉别人(当然,这些经历也常常被夸大)。

3. 过去分词分为规则和不规则变形,规则的变形都是动词原形+ed,与动词的过去式是一样的。如果动词的过去式是不规则变形,则它的过去分词通常也是不规则的。

原形	规则过去式	过去分词	原形	不规则过去式	不规则过去分词
enjoy	enjoyed	enjoyed	do	did	done
consult	consulted	consulted	take	took	taken
manage	managed	managed	become	became	become
focus	focused	focused	speak	spoke	spoken

 Ⅳ Exercise and Activity

1. 了解以下品牌，写下它们的成长过程。(What have they done back in the days?)

SIEMENS

novo nordisk

Microsoft

Apple

Disney

2. 与同伴进行一段以I have never done...或者 I haven't done... 为开头的对话，并询问对方"Have you ever done it?"

A: I have never been to the Shanghai Semi-conductor Compound. Have you been there before?

B: Actually I have been there last summer.

A: Great! Tell me about it!

I have never passed the ____________ test.

I haven't been to ____________.

I have never tried to ____________.

My company haven't ____________ yet.

Chapter C Describing a Project 描述项目

Ⅰ Key Words and Expressions

new organic 新有机物	throwing a party 开派对	youngsters 年轻人	issues 事务
gain great praise 获得赞许	retire 退休	willing to do something 决意做某事	in recognition of 认可
delivering 提供	tribute 奖励，贡献	crash test 碰撞测试	surveyor 测量员
facility management 设施管理	compound 工厂	analyse 分析	proceed 进行
chief technician 总技工	all-round 全方位的	pharmaceutical 医药的	equipment 设备

Ⅱ Read the Material Below and Answer the Questions

1. The CST's new organic compound: it's located in Chengdu. Our company completed this project in 2014 and have gained considerable praise from our clients.

We have also been delivering the facility management services since 2015, and CST is going to sign another project with us in Qingdao.

2. Jeff has been working in this company for the past 26 years and has been our chief technician for 12 years. We are throwing him a party today as he is going to retire from CCOC. Before he left, CCOC gave him 1 million dollars in recognition of his 26 years of service.

"I will always remember the good days in CCOC, and I'm willing to support the youngsters whenever they need it," said Jeff.

3. On the 26th of December, the quality control department had a meeting to enhance the Roche project's quality control issues.

They have done 5 all-around inspections towards the installation since last November and did another crash test this Monday. During the meeting, the chief engineer Steve has analysed the quality risk on the project and is going to proceed with a different plan to find a new solution.

Question

1. What services have we been delivering to CST's compound since 2015?
2. Why did CCOC give Old Jeff 1 million dollars?
3. How many inspections have they done since last November?

Ⅲ Core Knowledge of Describing in Different Directions

一般来说，作为用惯了一般式的中国人，我们能掌握的介绍事物的方式也许仅限于 He is Jeff. 或 This is the SIEMENS project. 但是学习了那么多的语法之后，我们应当尝试用各种不同的时态从各个方面来描述一个事物。

在描述一件事物前可以先列出一个表，将各个方面的信息填入，然后描述出一个较为完整的画面。

	Truth	Continuous	Future	Experience	Past
SIEMENS					
The ______________ Project					
Our Boss					

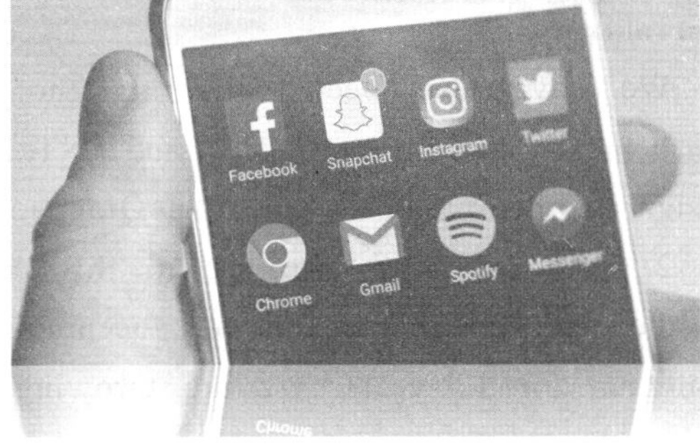

E.g. The surveyor Conan

Truth: His name is Conan. He is our company's surveyor.

Continuous: He is surveying the site now.

Experience: He has been doing this work for the past 5 years, and has done surveys for 9 projects.

Plan: He is going to send the final report next week to the headquarter.

Ⅳ Exercise and Activity

What is it? What do you think about it?

Anything that is going on now to it?

Think about the recent events of this product.

List the great achievements of this product.

Discuss your recent works with your partner, using these elements on the left. (请用多种时态进行描述)

What are you working on recently?	Well, we have been working on the firefighting measurement, and have done 5 inspections on site. Our manager is going to purchase another set of firefighting equipment next week.
Did you do anything last week for your work?	
Are you going to make any improvement in the project?	
Have you experienced any accidents on the site?	
Did you talk to anyone important last month?	
Who is the owner of your current project?	
What's your goal for the next 3 years?	

(趁热打铁,课后练习见143页)

Appendix Designer's Expressions 设计常用语

1. Useful expressions and sentences for designer's work.

We completed the task according to the drawing SF-1 last weekend. (completed the task 完成了任务)

You could have a look at the plan of our building. (have a look at the plan 请看看我们的蓝图)

This is the general view of this building. (general view 全视图)

How many drawings are there in the set? Moreover, what is the edition of the

drawing? (edition 版本; drawing 图纸)

Will there be any modifications to the drawing? (modification 修订)

Those pictures are too blurry; please send me another one. (blurry 模糊)

Could you please copy these drawings? (copy these drawings 复印图纸)

Can you explain the meaning of this symbol (abbreviation) on the drawing?(能给我们解释一下图纸上这个符号或缩写的意义吗?)

2. Useful words.

Floor Plan 平面布置	Directions/Positions 方向/位置		Drawing Information 图纸信息
general arrangement 总体安排	front 前/back 后	vertical 垂直	instruction book 说明书
details 细节	rear 背面	bottom 底部	operation manual 操作说明
section 区域	side 边	elevation 拔升	drawing number, size, scale, weight 图纸号码，尺码，大小，重量
installation 安装	left 左	auxiliary 备用的	drawing sketch 草图
civil construction 土建	right 右	cut-away 切线	technical specification 技术细则
general layout 总体布局	top 上/under 下	birds-eye 鸟瞰	diagram 图样

3. 下面的材料为项目执行计划，请用自己的话把它陈述出来。

Construction Progress Details	
Progress	execute
Fireproof Steel Beam Planning	OK
PVC	OK
Electrical Construction Plan	OK
Generatrix Improvement Plan	2 days ahead
Generatrix Modification Drawing	14 hours ahead
Re-examine the Water Pump Data	40%
DCC Contractor's Construction Statistics	100%
The Smoke-control Pipe Planning and Submitting	100%

4. Daily schedule (check list).

Week 1 (May 3rd)	Sales Tasks
Monday	Contact Mr. Sun before he leaves Shanghai.
Tuesday	Talk to the suppliers from ATPM and get a list of price.
Thursday	Confirm the delivery date of our steel structure vendors.
Friday	Train the EHS engineers on site.
Saturday	None.
Sunday	Lunch with Mr. Ryan.

Answers

Chapter A Report

Ⅱ Read the Material Below and Answer the Questions

1. They were discussing about the situations on the site (including staff and works).

2. She is the new technician. She was awarded the best engineer of the year because of her outstanding works, and she transferred to the northern branch this May.

Ⅳ Exercise and Activity

1. 将以下动词转换为过去式并翻译。

launch	launched 启动	negotiate	negotiated 谈判
keep	kept 保持	solve	solved 解决
cooperate	cooperated 合作	attend	attended 参与
manage	managed 管理	arrange	arranged 安排
loss	lost 失去	organize	organized 组织
win	won 赢得	engage	engaged 约定
control	controlled 控制	permit	permitted 允许
spend	spent 消费	promote	promoted 升职
talk	talked 谈论	document	documented 记录
communicate	communicated 交流	deal	dealt 处理

2. 用一般过去时回答以下问题。

Questions	Answers
Where was your last project located?	My last project was located in Shanghai.
Where did you work last week?	I was working in the CABOT factory last week, and I learned a lot.
Who was your first boss? Did you like him?	My first boss was Mr. Shiraz from Japan; I didn't really like him, because he was too bossy and mean!
When did you get off work yesterday?	I got off work late last night. It was about 3 a.m. when I arrived home.
What major did you choose in school?	I chose the information technology major. It was a promising (有前途的) major.
Who was your favorite project manager?	My favorite project manager was Ian Dawson. He managed the P&G project in 2013.

3. 阅读下文,回答以下问题。

the 21st of September	We finished our engineering training.
the 22nd of September	Our chairman paid a visit to us and did a presentation.
the meeting	The profect manager was with our clients from the USA.
the evening of 22nd of September	We had hot pot with our clients.

Chapter B Experience and Achievements

Ⅱ Read the Material Below and Answer the Questions

1. CCOC has been focused on and specialized in the electronic and pharmaceutical field since 2003. It has worked for famous companies such as the SIEMENS, CORNING and Microsoft, and has done projects such as the building of the SIEMENS Shanghai factory, the R&D building for BOE, and the research tower for the IVO Company.

2. Yes, he is. Because excellency has always been the core of CCOC's practice ever since the first day it was founded.

Chapter C Describing a Project

Ⅱ Read the Material Below and Answer the Questions

1. Delivering the facility management services since 2015.

2. CCOC gave him 1 million dollars yesterday as a special tribute for his 26 years of service.

3. They have done 5 all-around inspections towards the installation since last November.

Ⅳ Exercise and Activity

What are you working on recently?	Well, we have been working on the firefighting measurement, and have done 5 inspections on site. Our manager is going to purchase another set of firefighting equipment next week.
Did you do anything last week for your work?	I visited some important clients last week.
Are you going to do any improvements in the project?	I'm going to improve the ventilation system of our project.
Have you experienced any accidents on the site?	3 of our staffs have been injured by a steel external signage.
Did you talk to anyone important last month?	I talked to our president Mr. Carlson last Saturday.
Who is the owner of your current project?	Peter is the owner of my current project.
What's your goal for the next 3 years?	I'm going to greatly increase our workers welfares for the next 3 years.

Class 3

Opinions to the Case

看法

Chapter A This Is Our Team 我们的团队

Ⅰ Key Words and Expressions

meet at last 终于碰到了	professional 专业的	have great potential 有潜能	resourceful 资源丰富的,多才多艺的
Pleasure is mine. 我也很荣幸。	challenging 有挑战的	count on someone 依靠某人	competitive 好胜的
consist of 包含	supervise 监督	strict attitude 态度严格	practical 实干的
semi-conductor 半导体	decisive 有决断的	expect 期待	confident 自信的
extremely 极其地	careful 小心的	Job well done. 干得好。	mature 成熟的
keen 锐利的	sharing 分享	reliable 可靠的,靠谱的	honest 诚实的

Ⅱ Read the Material Below and Answer the Questions

Carl: Hello James, it's wonderful we meet at last!

James: It indeed is, and it's a pleasure for me to introduce our project team formally.

Carl: The pleasure is mine! Please go ahead.

James: Our team consists of 35 excellent engineers who specialize in the semi-conductor field.

Carl: Brilliant. Tell me more about them. May I ask, who is this Bryant Lee?

James: As you can see in the files, Bryant Lee is our project manager. If you ask me, Bryant is an enthusiastic and professional man. He has been working in this industry for the past 12 years and has managed some of the most famous and challenging projects we've had in the last 5 years.

Carl: Excellent. How about Ted Olsen who is in charge of safety, right?

James: That's right. Ted has been supervising our company's EHS issues since 2001. He is a careful but decisive man. Everyone calls him "Old Teddy" because he has many experiences in dealing with all kinds of incidents and he also likes to share his experience with our new recruits.

Carl: Haha, I'm looking forward to meeting him. What about this young lady Betty Davis, what's her job?

James: Well, Betty takes care of procurement. She is a procurement manager. She is entirely new to the business, but she has great potential and is hardworking. You can always count on her.

Carl: Understood. Who is this gentleman in the blue shirt?

James: You mean Stone? He is the technical expert who graduated from the

Northern Institute of Technology and has been focusing on the development of the semi-conductor field for the past 6 years. I think he is a man with strict attitudes towards work. You can trust him with anything.

Carl: Thank you, James! I will leave the rest of these files to my assistant. It's such an excellent team you've built up here. I'm expecting a job well done from them.

James: You will certainly get one, Mr. Carl.

Questions

1. What is this conversation about?
2. What does James think of Ted Olsen?
3. Why should Carl trust Stone?

Ⅲ Core Knowledge of Using Adjectives to Describe

1. Different kinds of adjectives.

Opinion	Size	Color	Shape	Nationality	Materials
excellent	large	black	round	Canadian	iron
professional	huge	red	square	American	steel
strict	tiny	green	rectangular	Japanese	PVC
down to earth	small	yellow	triangular	German	concrete

Tip：在用一个以上的形容词来形容一个事物时要注意形容词的排列顺序，即看法→大小→颜色→形状→产地→材质。

E.g. This is an excellent, large, black, round, Canadian, iron suitcase.

用上面的排列方法来描述一下这个Rolex手表。

2. Opinion adjectives.

Positive Adjectives				Negative Adjectives	
talented	reliable	honest	proud	bossy	arrogant
creative	resourceful	mature	influential	demanding	selfish
practical	respectful	experienced	popular	impulsive	disorganized
confident	impressive	optimistic	famous	aggressive	hateful
traditional	generous	competitive	responsible	unprofessional	cheap

Tip 1：这些关于看法的形容词也可以用来描述组织、公司、团队或者产品，如果能顺带加上because...来支持你的观点的话就更好了。

really	pretty
extremely	not at all
absolutely	a little bit

Tip 2：程度副词如really，extremely等能用来强化形容词。而kind of，not at all等可以用来弱化形容词。

Ⅳ Exercise and Activity

1. 用以下形容词重写句子,让它们具有相反的意思。

experienced	untrustworthy	easygoing	cheap	polite	unknown

My project manager is really demanding!	My project manager is pretty easygoing.
Our rival's（竞争对手）product is really expensive.	
My colleagues are pretty rude to me.	
Bobby was very new to his job.	
Dr. Stevenson is quite famous in the chemical field.	
CCOC is a reliable facility management company.	

2. 用看法类形容词描述以下人和物,并解释为什么你这样觉得。

your boss	
your department	
your favorite company	
a book you like to read	
a project you are doing	

Chapter B Relationships 关系

Ⅰ Key Words and Expressions

committee 委员会	counseling 咨询	hopefully 但愿	responsible 负责的
authorization 授权	interview 面试	bargains 减价品	influential 有影响力的
cousin 表亲	bring into 带入行	lubricant oil 润滑油	convention 会议
legal guidance and support 法律指导和支持	husband/wife 丈夫/妻子	It's both...and... 不但……而且……	famous 有名的
pay a visit 拜访	multi million company 数百万市值的大公司	childhood 童年	generous 慷慨的

Ⅱ Read the Material Below and Answer the Questions

1. Caroline Joseph Lin is a critical member of the committee that supervises the authorization of district 27's construction projects. She is a good friend of our rival company's CEO Mr. Anderson, who graduated from the same college as she did. You

should definitely speak to her before our company starts this project.

2. Rowland Kenneth is the district attorney (检察官) who is also the cousin of our boss. He is the most influential man in town, and he could provide us with legal guidance and support. Why don't we pay him a visit after work?

3. Simon Lin is an energetic and reliable man who works in the same project with me. He is a guy that I interviewed and brought into our department. I think he is the husband of Caroline who works in the committee of district 27.

4. Tony Peterson is someone I spent most of my childhood with. He is almost like a big brother to me. Now he is the boss of a multi-million dollar company that supplies all kinds of construction materials. I will give him a call tonight to discuss future co-operation. Hopefully, we could get some good bargains from him.

Questions

1. Who could we speak to if we want to get a reach to Caroline?
2. Why would Tony give our company some discount?

Ⅲ Core Knowledge of Related Things Together in One Sentence

Who	Which	That
someone	something	someone
		something

Why do we use adjective clause?

He is a wonderful man.	He is a wonderful man who always helps his workmates. (two ideas combined)
He always helps his workmates.	

关系从句不仅将两个不同的观点合并在了一起,同时也为听众提供了准确的指向。

 Ⅳ Exercise and Activity

1. 请用关系代词(who/which/that)连接以下句子。

I have a friend. She works in lubricant oil company. She works really hard to support her family.

I meet this Brazilian guy from this bio-pharmaceutical convention. He organizes a lot of parties for pharmaceutical experts.

This is my closest friend Kim. He owns a chemical company.

2. I ask, your answer.

Who is your first boss? How did you meet?	
Do you have any clients you met through other friends?	
Do you keep in touch with your old friends?	
What's your company? What's its business?	
What's your favorite job? What is it about?	
Tell us about a client that you dislike (不喜欢).	

Chapter C It Seems to Me 在我看来

I Key Words and Expressions

under great pressure 在巨大压力下	supplementary provisions 补充条款	relationships 关系
be approved 通过了	as a matter of fact 事实上	registration 注册
law statements 法律条款	exam 测试	submissions 提交
general provision 总则	purpose 目的	government 政府
building permit 建筑许可	enhance 强化	production cost 生产开支
legal liability 法律责任	maintaining orders 保证纪律	show off 炫耀

Ⅱ Read the Material Below and Answer the Questions

Reece: So, how are your projects going?

Connie: Not bad, but our engineers are under enormous pressure because our authorizations and permits are still not approved yet.

Reece: It seems that there have been many changes to the laws and regulations in the last few years, which might be the reason why you can't get them very fast.

Connie: Well, I don't know much about construction law in China. Can you briefly explain the key issues to me?

Reece: There are 8 main chapters, which are: general provision, building permits, contract issuance, contracting of construction projects, and the supervision of construction projects. Let's see what else...ah! Yes! The safety management, the quality control of construction project, legal liability, and supplementary provisions.

Connie: Wow! How did you remember that?

Reece: As a matter of fact, I have just finished my exam on construction laws.

Connie: Cool! So I guess you could tell me about the purpose of the

construction law.

Reece: If you ask me, I think the law is to enhance the supervision and administration over building operations and maintain order in the construction markets. Also, of course, to guarantee the safety and quality of a project.

Connie: I don't even understand half of what you are talking about.

Questions

1. Why is the construction authorization not approved yet?
2. What's the purpose of the construction law?

Ⅲ The Core Knowledge of Giving an Opinion

Straight Forward Opinions	Add These Expressions to Make It Personal	
Our technician is too lazy!	I think that	If you ask me that
Too many young people nowadays are unemployed.	I don't think that	It seems that
Our owner hates our job!	I believe that	It looks like that
The local government created many problems for us.	It seems to me that	To be honest that
当我们用这种句式来陈述想法的时候,往往会让对方感觉过于直接,因为我们的想法也并不一定是事实。	将以上的表达句式放在观点之前会使语句婉转,并且仅代表个人观点。	

Rewrite these sentences above, and give us your opinions.

I don't think our technician is too lazy. I believe that he just doesn't like to show off his work.

Ⅳ Exercise and Activity

组成二人小组讨论是否支持以下话题,并给出理由(不少于3个理由)。

1. Good service means high price.
2. Is the relationship with the government the most important part of our business?
3. Should we put cameras to monitor everyday work?
4. Do we need to spend money and time on training?
5. What do you think about the term "the cheapest wins the bid"?

(趁热打铁,课后练习见144页)

Appendix Permit Submission 申请许可证

1. Permitting process for new building（建设新设施）.
planning registration 申请计划（provided by landlord）
plan submission and approval 计划提交并审核(provided by landlord)
file submission and approval（7 days）相关文件提交审核
environmental submission and approval 环境状况提交审核
all construction drawing 提供图纸
plan permission（20 days）计划批准
bidding process（30 days）竞标过程
construction permit（30 days）建筑许可
power supply 供电
water supply 供水
sewer/storm drains 下水道系统
natural gas 天然气
telecom 通信
acceptance（10 days）申请通过
2. Permitting process for the renovation of building（翻新设施）.
planning registration 申请计划（provided by landlord）
fire submission and approval（7 days）防火申请审核
environmental submission and approval 环境状况提交审核
all construction drawing 提供图纸
bidding process（30 days）竞标过程
construction permit（30 days）建筑许可
power supply permit 供电许可
water supply permit 供水许可
sewer/storm drains permit 下水道系统许可
natural gas permit 天然气许可
telecom permit 通信许可
acceptance（10 days）申请通过

Answers

Chapter A This Is Our Team

Ⅱ Read the Material Below and Answer the Questions

1. It's a brief introduction of James' company's project team.

2. Ted is really reliable because he is both experienced in dealing with all kinds of incidents, and is also very kind while sharing his experience with new recruits.

3. Stone has been focusing on the development of semi-conductor field for the past 6 years. He is a man with strict attitude towards work.

Ⅳ Exercise and Activity

1. 用以下形容词重写句子，让它们具有相反的意思。

My project manager is really demanding!	My project manager is pretty easygoing.
Our rival's（竞争对手）product is really expensive.	Our rival's product is really cheap.
My colleagues are pretty rude to me.	My colleagues are pretty polite to me.
Bobby was very new to his job.	Bobby was very experienced to his job.
Dr. Stevenson is quite famous in the chemical field.	Dr. Stevenson is quite unknown in the chemical field.
CCOC is a reliable facility management company.	CCOC is an unreliable facility management company.

2. 用看法类型形容词描述以下人和物，并解释为什么你这么觉得。

your boss	My boss is an outstanding manager in the construction industry.
your department	Our department is absolutely mature and experienced.
your favorite company	Google is my favorite company, because it's both creative and cool!
a book you like to read	*Human Nature* is the most fantastic book I have ever read.
a project you are doing	The KCS project is an outstanding project.

Chapter B Relationships

Ⅱ Read the Material Below and Answer the Questions

1. Simon is the man who is the husband of Caroline.

2. Because Tony is a friend who grew up with the speaker.

Ⅳ Exercise and Activity

1. 请用关系代词(who/which/that)连接以下句子。

I have a friend who works in a lubricant oil company that works really hard to support her family.

I meet this Brazilian guy from this bio-pharmaceutical convention who organizes a lot of parties for pharmaceutical experts.

This is my closest friend Kim who owns a chemical company.

2. I ask, you answer.

Who is your first boss? How did you meet?	My first boss was Mr. Dwight and we met in a sales conference.
Do you have any clients you met through other friends?	Dr. Milo is a client that my friend Peter introduced to me.
Do you keep in touch with your old friends?	Damian is my closest friend who I grew up with!
What's your company? What's its business?	CEFOC is a company that provides construction services.
What's your favorite job? What is it about?	My favorite job is to be a baseball coach that teaches teenagers.
Tell us about a client that you dislike (不喜欢).	TDSD's owner is an owner that I really don't like...

Chapter C It Seems to Me

Ⅱ Read the Material Below and Answer the Questions

1. Because many law statements have been changed recently.

2. It's to enhance the supervision and administration over building operations, and maintain order in the construction markets. Also, of course, to guarantee the safety and quality of a project.

Class 4

Engineering Presentations

工程类演讲

Chapter A Introducing Your Company 介绍你的公司

I Key Words and Expressions

greetings 招呼	growth 增长	project investment 项目投资	president总裁
representative 代表	functional sector业务单元	completion 完工	board leader (chairman) 董事长
I'm honored to... 荣幸	business sector 商务单元	frequent inspection 频繁检测	finance department 财务部
specialize 专门	sub-branches 分公司	strict standards 严格标准	marketing department 市场部
speech 演讲	EPC/EPCM 设计采购施工/设计采购和施工管理	supervision 监管	certain field 某个领域
background 背景	walk through 带领	organization chart 组织架构	annual revenue 年利润

II Read the Material Below and Answer the Questions

1. The front page (首页).

The main page of your PPT is the part where you say hello and introduce yourself to your clients. 在第一页我们要先介绍自己和今天需要讨论的话题。

Greetings (formal)	Who You Are	Your Job	Your Purpose
Morning L&G	I'm + name	The representative	I'm going to deliver...
L&G How are you		The project director	I'm honored to give you a speech about...
Greetings L&G	You can call me + name	The project manager	
		The EHS manager	

2. The contents (目录).

We will look into items from who we are to our goals.

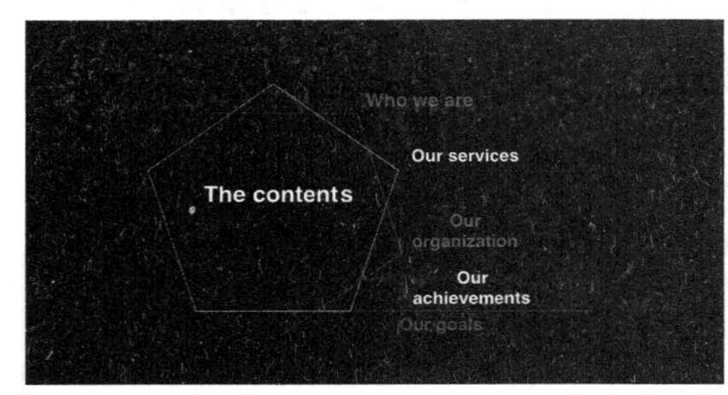

Who We Are	The Services	Organization Chart	Achievements	Goals and Visions
background	chemical	CEO	brief achievements	EHS
growth	electronic	functional	typical details	services
qualifications	pharmaceutical	business		business
	manufacture	branches		why choose us

3. Who we are, our background (我们是谁,我们的背景).

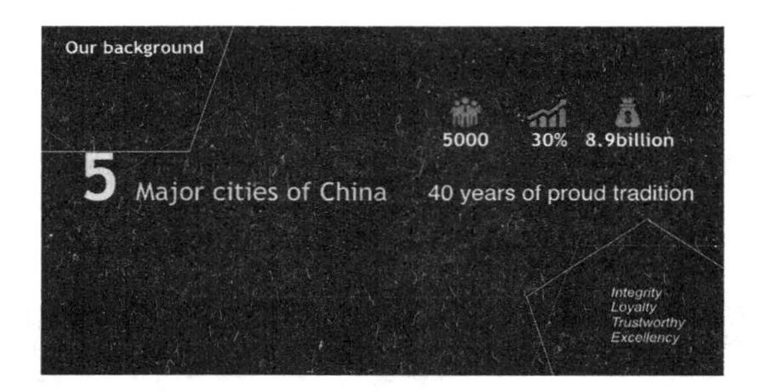

(1) In this page, you should explain all the important items that are shown on the page. But the most important of all is to tell your clients about your opinion towards your company.

our office coverage	The company has covered ____________ major cities.
our employees	We have a total of ____________ employees.
how much money we made	Last year we have reached ____________ of contract value.
what you think about your company	I believe that our company is a ______________________.

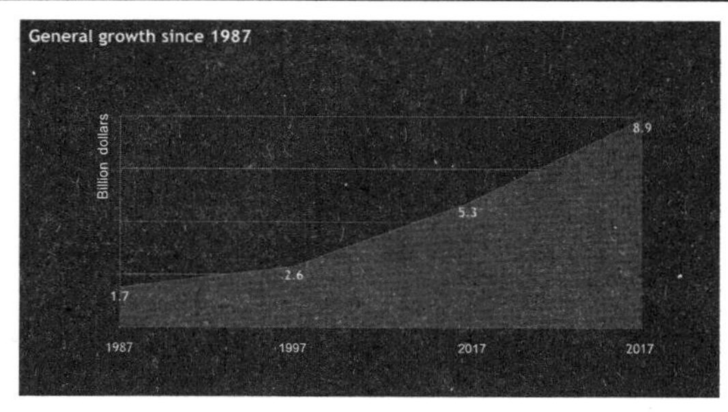

(2) Second, the history of the company's growth.

(3) Third, the data of your company's history.

Past	Achievement	Future
the beginning of the data	what you have done during the years	your future goals to grow in the future

(4) Briefly go through the qualifications and awards of your company.

(5) We have obtained qualifications from...

4. Our services (我们的服务).

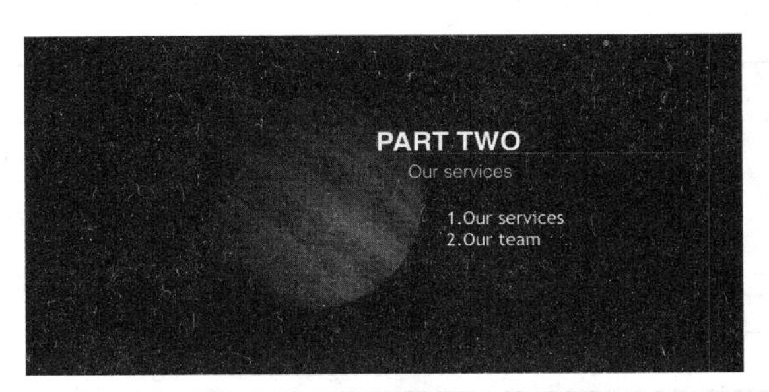

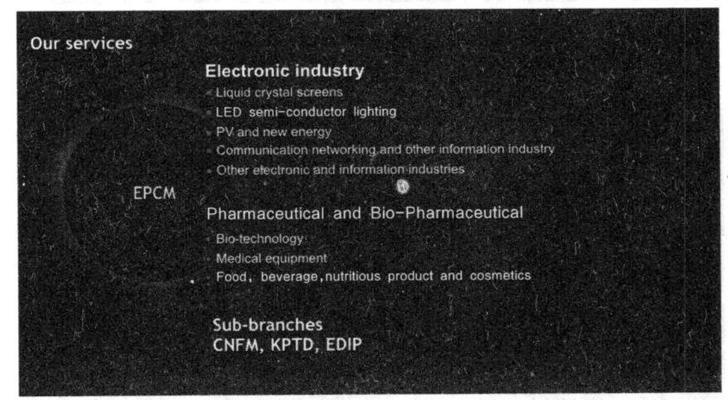

the company's main services	EPC, EPCM
fields we cover	Electronic, bio-pharmaceutical, manufacture and chemical industry
our team	That specializes in...
what you think about your services	It seems to me...
service details	...

Tip：可以根据客户所从事的产业来细致描述针对某个领域的服务。

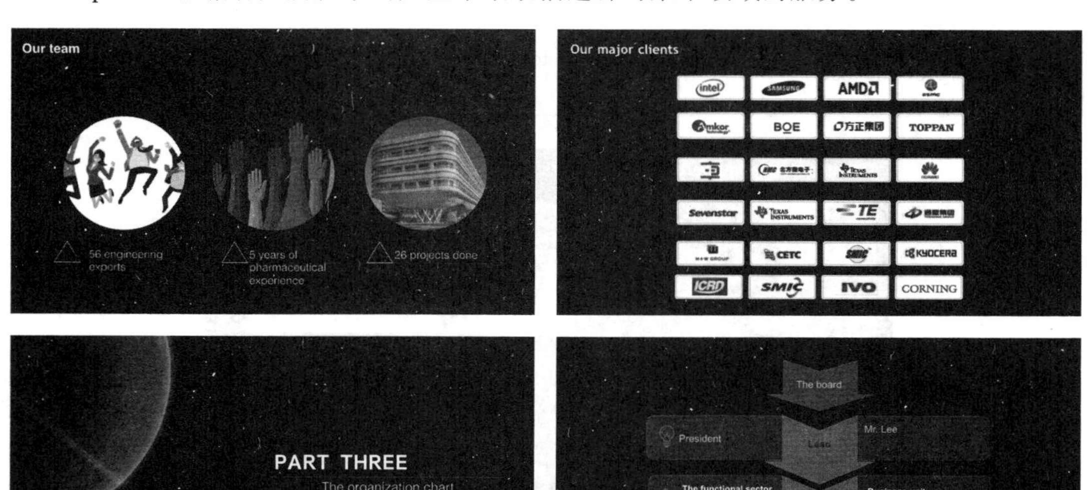

(You can also promote your team.)

5. The organization chart (组织架构).

introduce the board leader	Mr. Johnston is our board leader (chairman).
introduce the president	Our president is Mr. Lee who is a... and has done...
the functional sector	There are 13 departments including the HR, QC departments... (It supports...)
the business sector	The operation unit and the sales unit
the sub-branch services	From manufacture, designing, to engineering and facility management

6. Achievements and experiences (公司业绩).

The first page is a brief walk through of the company's past experiences in a certain field.

The second page is the detailed project achievements.

project owner	________ company
project location	It is located in ________.
total investment	The total investment is about ________.
project completion	It was completed in ________.（如果项目仍在进行则It's going to be completed in ________.）
services	Our services includes...

7. Goals and codes（最后一页，目标和原则）.

EHS Goals	Management Goals	Business Goals	Code of Practice
zero incident	strict standards	annual revenue	integrity
frequent inspection	intense supervision	international vision	loyalty
summarize	be more efficient	new branches	trustworthy
		new services	excellency

Ⅲ Easy Tips of Making a PPT Presentation

1. 不要读自己的PPT，多告诉客户他们所不知道的信息，多将你的看法讲述给他们听。

2. 每翻一页PPT都要告诉听众这一页的主旨，例如：So, this page is about our services. All right, the next page is about...

3. 当排列的服务、条例或者项目较多的时候，可以用“from...to...”来做一个概述，不需要一个个项目读出来。

4. 当图标和图片较多的时候，可以用“you could have a look”让听众自行观看。

5. 被问到较为深入的问题时，可以用“Good question. We would be happy to show you the details after the speech.”这样来让对方把问题放到提问阶段。

6. 在结束的时候，不要忘记说“thank you”。

Ⅳ Exercise and Activity

组成四人小组,集思广益,填写下方表格,并以此为基础,创建一个属于自己小组的公司介绍模板。最后,将公司介绍以演讲的方式展示出来。

name of the company	DP&P's Engineering Consultancy
what it does	consultancy
the company's background	
the company's growth	
the company's qualifications	
your company's achievements	
the goals of your company	

Chapter B Project Execution Plans 项目执行计划

Ⅰ Key Words and Expressions

kick off 启动	schedule control 日程管控	suppliers 供货商	rectification 更正
EHS control 安全管理	cost control 成本控制	local resources 本地资源	summarize 总结
quality control 质量管理	risk control 风险控制	principal 主要分析	guarantee 保障
BIM 建筑信息模型	documentation 文件管理	set up a team 成立团队	efficiency 效率
procurement management 采购管理	acceptance 接收	take actions and measures 做出行动	space management 空间管理
most suitable goods 最合适的产品	third party 第三方	sample 样本	discounts 优惠
centralized procurement 集中采购	Let's take a short break. 休息片刻。	take it from here 让某人从这接手	large and special equipment 大型特殊设备

Ⅱ Read the Material Below and Answer the Questions

Brandon: So, you are the manager of the RTS project right?

Todd: Yes sir. It's an honor to be the manager of this project. I'm Todd Jackson, pleasure to meet you.

Brandon: The pleasure is mine. You are here to tell us about the project execution plans, correct?

Todd: That's right. As you can see from this page, the plan covered services from the project kick-off to the acceptance of the project. Also, I will be RTS's project manager, and I have already set up a team that specializes in pharmaceutical projects. The suppliers are ready to deliver high-quality materials to us.

Brandon: Fantastic. Go on, please.

Todd: This page is about our EHS procedures, from EHS training to EHS inspections. If we find any problems, we will take actions and measures immediately and follow up with the appropriate changes.

Brandon: I believe your company also provides brilliant EHS training.

Todd: That's right, and our EHS and quality training is pretty well known because of the strict standards. Oh, and in this page here is our QC procedure. It's the same as the EHS procedure, to guarantee our quality we also provide BIM services to our clients.

Brandon: BIM? You mean the building information model?

Todd: That's right. It gives us efficient space management, installation set up, and most important of all, a beforehand prevention on the screen before anything happens.

Brandon: The power of technology! Let's have a look at your procurement management.

Todd: Sure. We have got an experienced team that can select the most suitable goods for you. Also, to ensure the high performance of the materials, we would send our samples to a third party and provide you with an additional sample for examination. Large and special equipment is not included, however.

Brandon: So, what does centralizing procurement mean?

Todd: Well, it means that if we purchase plenty of materials from the same supplier and use them on different projects, the more we buy, the more significant the discounts. Also, this page is our schedule control. Here you can see some forms for our project schedule updates for the first 21 days.

Brandon: I see...Um...Why don't we take a short break so that we could look into it?

Todd: Sure thing sir, my colleague Michael will take it from here.

Brandon: Well done, thank you.

Questions

1. How many items did Todd introduce?
2. Why is Todd's company's training very well known?
3. Please explain the meaning of centralized procurement.

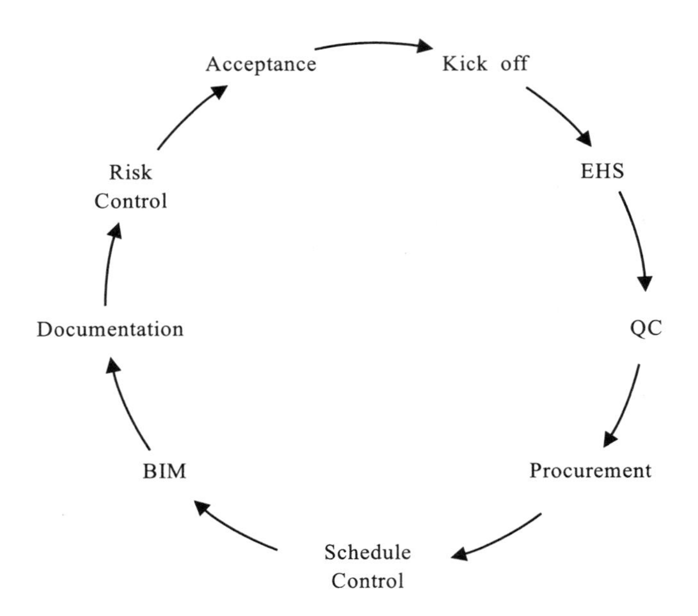

Ⅲ Tips of Making a PEP Speech

1. The kick off（项目启动）.

Your project manager	___________ is going to be your project manager, and he has done _____ _____ projects.
Your team	The team we set up specializes in ___________.
Your suppliers	Our suppliers delivers excellent materials...
Your local resources	We maintain wonderful relationships with the local government.

2. EHS（安全）.

The EHS Procedure	The EHS Training	EHS Inspections	EHS Meetings
From EHS training to EHS inspections, if we find problems, we would...	training acceptance	daily, weekly, monthly	daily meeting schedule
	training management	other inspections	summarize
	pass the exam	rewards and punishments	rectifications
	authorization card		rewards and punishments

3. QC（质量监管）.

The QC Procedure	Keys of QC	QC Inspection
From QC training to QC inspections, if we find problems, we would...(basically the same as the EHS procedure)	the quality of main structure	QC training
	pressure test	QC meeting
	the safety of special equipment	Make sure the installation is right.
		compensations（赔偿）
For detailed QC issues, you could refer to the QC SOP books.		

4. BIM（建筑信息模型）

What is BIM?	BIM = building information model (an extra service from our company)
What does it do?	efficient space management, prevention, installation
the principal of BIM	to create a 3D visual environment to help us find problems beforehand
the BIM walks through	talk about the items from the image

5. Procurement（采购）.

the procurement process	from initiating the procurement to updating the procurement
samples for examination	provide samples for the third party and party B
centralized procurement	the larger the purchase the larger the discount
procurement update	update the procurement list and follow up the receiving date
procurement team/graph	introduce the procurement team and the graphs in the past

6. Schedule control（日程管理）.

schedule control process	From this page you could see the items of our schedule control process, from our milestone to daily updates.
process editing	We will edit our process due to (1) the date of receiving the materials, (2) different phase of each sector's work, (3) HR organization, and (4) testing plans.
updating schedule	from 21 days of updating plans to daily plans
feedbacks and corrections	We would gather feedbacks from daily plans to weekly plans, and make rectifications and updates for those plans.

7. Documentation（文件管理）.

documentation	Our documents are kept safely in our date center.
document preservation	3 years of preservation, all files, all testing samples

8. Risk control（风险控制）.

risk identification	operation risks, and building system risks
risk inspection and investigation	The ________ (risk) was mainly caused by ________.
solution	We would tackle this problem by ________.

9. Hand over（项目交接）.

handover	the handover process
custom service	sign and perform the quality guarantee agreement

Ⅳ Exercise and Activity

与同伴互相问答以下问题。

Could you please tell us about your training policy towards EHS and QC?
How do you cope with pipe leakage?
What is centralized procurement?
What's the strength of BIM?

Chapter C Support Your Opinions 支持观点

Ⅰ Key Words and Expressions

campus 产业园	underestimate 低估	dead end 死路	overconfident 过于自信
strength/weakness 优/劣势	awful disaster 大灾难	in addition 还有	connection 关系
no wonder 怪不得	terrible 糟糕	local clients 本地客户	as for conclusion 关于结论
without hesitation 毫不犹豫	cheer up 加油	foreign clients 国外客户	good enough 足够好
impress 留下印象	globalized 全球化	coming up 即将到来	basically 基本上

Ⅱ Read the Material Below and Answer the Questions

Walker: As far as I know, we have won the bid for the R&D campus of AZT.

Kim: That's right. It's quite a project for our company.

Walker: So what are our strengths that impress AZT?

Kim: Well, if you ask me, first we provided fantastic before—construction services like the BIM services and excellent design drawings. Second, I think our brand also helped us a great deal in the bidding process.

Walker: No wonder AZT selected CCOC without any hesitation.

Kim: Yea, It's going to be much work coming up this July. So, what about the bid you guys are doing? Is it the project for Intel?

Walker: Ah...We are not doing so well. Our rival company CNCEC has won the project.

Kim: How come? What happened? I thought you guys were pretty confident last week. How did you lose the bid?

Walker: I think that we are sometimes overconfident, and we underestimated CNCEC's connection with Intel. What's more, we were not ready for the bid—our presentation was a disaster. Besides, our manager's English was just terrible.

Kim: Oh well, cheer up buddy. There is another bid coming up on Friday this week right? Just stay focused!

Questions

1. Why did Kim's team win the bid?
2. Why didn't Walker's team win the bid?

Ⅲ How to Support Your Own Opinions

给出建议	It seems to me... If you ask me...
表明观点	Basically I believe... The thing is... You see...
列举论点	First, second, third... One, two, three... A, B, C...
提供解释	It's because... It's probably because... It's caused by...
补充	On the other hand... And another thing is... In addition... For example...
结论	In conclusion... Therefore...

Tip: You can tell a story or give an example to support your views.

Agree with the View 同意观点的回应	Disagree with the View 不同意观点的回应
I think you are right.	I don't know about that.
That's true.	It's true, but...
Well, I'm with you.	Right. However...
For sure.	I disagree with your view because...

Ⅳ Exercise and Activity

组成小组,针对下列观点进行辩论。

We should find more international projects. 支持	We don't need to find too much international projects. 反对
In a globalized world, the outside world has more opportunities because...	We have enough market in China already. We don't need to go outside.
It's a great chance for us to present our company to the world market.	It will waste a great amount of resources doing this.
We are going to learn more about this industry by working with the foreigners.	We are good enough for our local clients; we should be focused in our own ground.
If we don't make a change, we would meet a dead end.	Making a change means risk because...

(趁热打铁,课后练习见145页)

Appendix Bid Forms 标书

To (name of the bidding agent)

In accordance with your IFB NO ____________ for (goods/services supplied) for the ____________ project, the undersigned representative ____________ duly authorized to act in the name and the account of the bidder (the name and address of the bidder) hereby submit the following in one original copies.

Bidding Form Items

summary sheet for bid opening 开标一览表	qualification documents 资格证明文件
bid schedule for prices 投标分项标价表	Bid security amount __________ issued by name of the bank 投标保证金保函,金额______
brief descriptions of the goods/services 货物/服务说明一览表	The bid is valid for a period of ____________ calendar days. 本投标有效期为____________。
responsiveness/deviation form for commercial terms 技术规格响应/偏离表	The bidder agrees to the provision about bid security forfeiture. 同意投标人须知中关于没收投标保证金的规定。
all the other documents required in response to instruction to bidders and technical specification 投标人须知和技术要求提供的有关文件	To declare that the bidder is not associated with a firm engaged by the tendering agent/the purchaser to provide no services for this project or working for the firm. 承诺与买方任何附属机构无关,也不属于买方附属机构。
The bidder agrees to furnish any other data or information pertinent to its bid that might be requested by the tendering agent. 提供任何与投标相关的文件与信息。	All correspondence pertinent to this bid should be addressed to... 与本投标相关的一切正式信函请寄至……

Name of the Representative ____________________ 代表人姓名

Name of the Bidder ____________________ 投标人姓名

Official Seal ____________________ 公章

Date ____________________ 日期

Answers

Chapter A Introducing Your Company

Ⅳ Exercise and Activity

组成四人小组，集思广益，填写下方表格，并以此为基础，创建一个属于自己小组的公司介绍模板。最后，将公司介绍以演讲的方式展示出来。

name of the company	DP&P's Engineering Consultancy
what it does	consultancy
the company's background	It has been providing consultancy services to major engineering firms worldwide.
the company's growth	The company promises 30% revenue growth each year.
the company's qualifications	The company has obtained qualifications of class A.
your company's achievements	We have successfully delivered excellent consultancy services to companies such as... Here I would like to give you some details of the work.
the goals of your company	We are going to...

Chapter B Project Execution Plans

Ⅱ Read the Material Below and Answer the Questions

1. Six items. They are project kick-off, EHS, QC, BIM, procurement management, and schedule control.

2. It's famous for its strict standards.

3. It means purchasing a large deal of materials from the same supplier to provide them into different projects. The larger amount we buy, the larger the discounts we will get.

Ⅳ Exercise and Activity

与同伴互相问答以下问题。

The EHS training and the QC training policy is almost the same, from compiling the training to authorizing a pass card.
The pipe leakage is mainly caused by... We would normally tackle this problem by...
It means to purchase a large deal of materials from the same supplier to provide them into different projects. The larger amount we buy, the larger the discounts we will get.
Effective space management, beforehand prevention, installation

Chapter C Support Your Opinions

Ⅱ Read the Material Below and Answer the Questions

1. First they provide fantastic pre-construction services like the BIM services and excellent design drawings. Second, they think that their brand also helps them a great deal in the bidding process.

2. One, they are too overconfident, and underestimates CNCEC's connection with Intel. And two, they are not really ready for the bid, and their presentation was an awful disaster. In addition, their manager's English was just terrible.

Ⅳ Exercise and Activity

组成小组,针对下列观点进行辩论。

In a globalized world, the outside world has more opportunities because...	We have enough market in China already. We don't need to go outside.
I agreed. Because firstly opening up the world market can give us a lot of new opportunities. Secondly...	It will waste a great amount of resources doing this.

(仅供参考)

Class 5

Problem Solving
解决问题

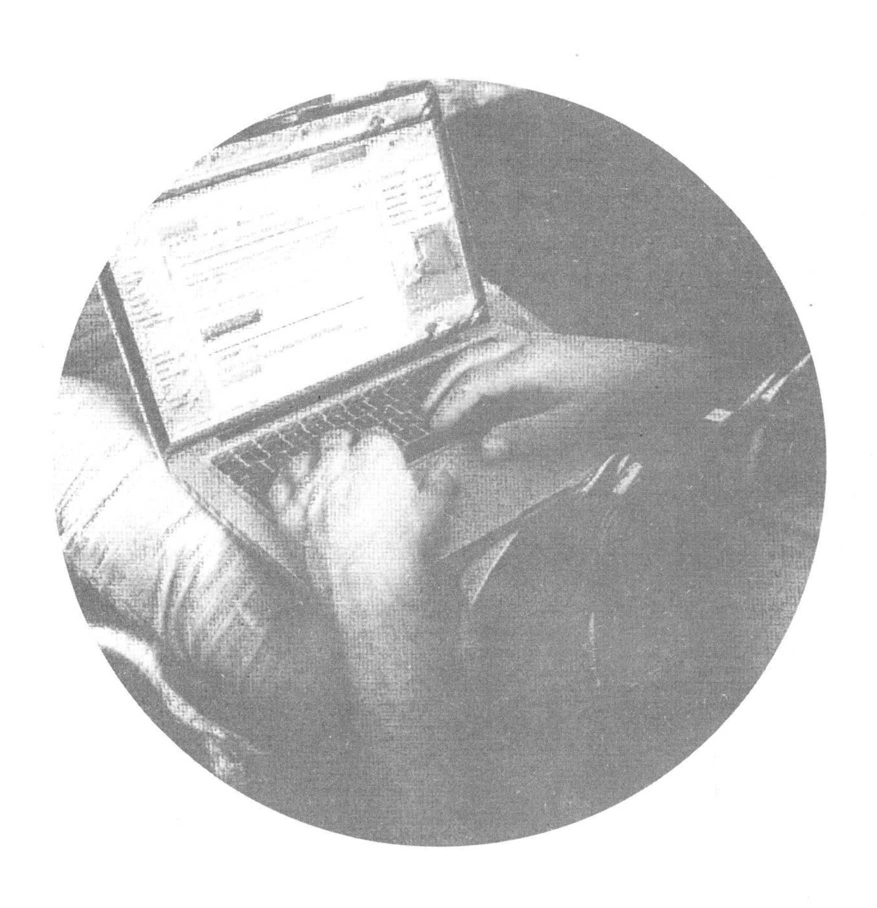

Chapter A Clarification Emails 澄清邮件

I Key Words and Expressions

concern 抱怨，意见	CV 简历	unfortunate 不幸的	wordiness 字数太多
embassy 使馆	impressed 印象深刻的	tackle the problem 解决麻烦	concrete language 确定的口吻，具体的语言
water leakage 漏水	lubricant oil 润滑油	understanding 理解	willing to learn 有意愿学习
take care of it 解决它	headquarter 总部	Would it be OK to... 可否……	influence someone 影响某人
flood the building 淹没建筑	hospitality 招待	manners 礼貌	cover it up 掩盖

Ⅱ Read the Material Below and Answer the Questions

To：steveloveparis@ccoc.cn
Subject：The Broken Pipe in the ROK Embassy
I'm writing this email to concern about the broken pipe on the first floor of the embassy. The water leakage seems to come from the cooling engine, so please, send some staffs to solve this problem as soon as possible before it floods the building.
Best Regards
Richard Madison

To：TimDawson@hotmail.com
Subject：Project Chief Technician Interview
We are writing this letter to thank you for sending us your CV, and we are impressed with your experiences in the lubricant oil industry as a chief technician. We would like to see you at 10:30 a.m. on January 19th, at our headquarter office in Berlin. Please confirm this letter by sending us a confirmation email or a phone call (Tel:39878527 or marianyang@ccoc.ger). Look forward to seeing you soon.
Best Regards
Marian Yang

To：petegriffinrules@ccoc.cn
Subject：The Delay of the Installation
This is a letter to acknowledge you the message about the seven days delay installation towards the TDSD R&D center's HVAC system. We were hoping you could provide us with a reasonable explanation or a solution to the case within 24 hours. Alternatively, we would report this to our headquarter. Thank you for your attention. Please take this into urgent consideration.
Best Regards
Mathew Billiton

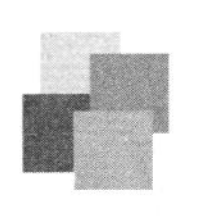

Questions

1. What does the term "flooded" means?
2. Why are Marian's company impressed with Tim's CV?
3. What does Mathew need from Peter's company?

Ⅲ Core Knowledge of Writing an Email to Solve Problems

1. 邮件分解。

Who to?	Dear Mr./Mrs./Ms...	To whom it may concern.	Hello Mr./Mrs./Miss...
What for?	The problem about...	Congratulations to...	The meeting.
Reasons of writing.	We are writing this to concern about...	We are writing this letter to apologize about...	We are writing this letter to ask for your support.
The cause of the problem.	After our investigation we found out that...	The problem was mainly caused by...	It's unfortunate that... happened.
The solution to this problem.	We are going to...	Tackle it.	Solve it.
Ask for permission.	Would it be OK?	Is it all right to...?	Would it be fine to do it?
Thanking words.	Thank you for your concern.	Thank you for calling.	Thank you for understanding.
Ending + the name of the sender.	Sincerely yours.	All the best.	Best regards.

2. 写邮件的简单技巧。

Mind your manners.
Write briefly, avoiding wordiness.
Use simple language.
Use concrete language like within 5 working days, within 24 hours or before June the 2nd.
Create a specific, meaningful and straightforward subject line.

Ⅳ Exercise and Activity

To: petegriffinrules@ccoc.cn
Subject: The Delay of the Installation
This is a letter to acknowledge you the message about the seven-day-delay of installation towards the TDSD R&D center's HVAC system. We were hoping you could provide us with a reasonable explanation or a solution to the case within 24 hours. Alternatively, we would report this to our headquarter. Thank you for your attention, and please lease take this into urgent consideration.
Best Regards
Mathew Billiton

1. 请按格式回复以下信件。

To:
Subject:

To: chrismogan2213@prado.us
Subject: The Budget Error of Procurement
We are writing this letter to inform you about the error of the procurement budget's invoice dated 21st of June 2017. It should indeed read as AUS$ 200,000 instead of AUS$20,000. Please make an immediate correction before next Monday. Thank you.
Best Wishes
Helen Montana

To:
Subject:

To: excecutiveoffice@ccoc.cn
Subject: Fire Billy Zhou
I'm upset about the hospitality of this project. I have explicitly asked for a 5-star hotel but what I get was just a 4-star, and Billy Zhou, the guy who arranged my stay, is an idiot! He said that the place we set up the SHELL project didn't have any 5-star hotels, and 4-star is the best he could get. So he went and booked me a 4-star hotel! I have called him several times to tell him that I'm someone important, but he just gave me a bunch of excuses! I want this guy to be fired ASAP! Or else I would take actions!
Yours
Reece_Luther@TPTA.com

To:
Subject:
The Executive Office

2. 请依据下列需求给以下人物(根据给予的关系条件)写一封电子邮件。

Your owner, Bill(very demanding)
Write a sick leave (请假条) for 3 days.
You are handing your job to Mark.
Mark is a new guy, but he is willing to learn.

Your boss, Paul (quite easygoing)
The project needs more workers.
The project was delayed.
You need to change the engineer Steve.

Maggie, the expert (quite helpful)
You need some help when translating the owner's needs.
You need her suggestion towards the newest HVAC system from Japan.
You want to ask her out for a lunch.

Bowen, your workmate and friend (very close)
You made a big mistake during the cleanroom construction, and it influenced Bowen.
You need him to help you cover it up.
You want to apologize to him for creating such trouble (麻烦).

Chapter B Helping Hand 寻求帮助

I Key Words and Expressions

get it fixed 修好它	power outage 漏电	crane operator 吊车司机	cost estimator/ engineer 造价员	exhaustion 疲惫
need to be repaired 需要修理	ceiling 吊顶	plumber 管道工	structure crew 施工员	talk through the matter 讨论解决
need replacing 需更换	quality inspector 质检员	driller 钻工	architect 建筑师	push over 推过来
have someone do something 让某人做某事	scaffold 脚手架	bull dozer operator 推土机司机	road builder 筑路工	concrete worker 水泥工
cut the power 断电	warehouse keeper 仓库管理员	steel fixer 钢筋工	under control 得到控制	

II Read the Material Below and Answer the Questions

Ken: Anna, did you hear the noise by the ceiling?

Anna: I didn't. Oh, wait, yes... it sounds like something is leaking out and dripping (滴水)!

Ken: Let me have a look. Could you push over the scaffold for me? I have to get it checked!

Anna: Sure, wait up. Here you go. Be careful.

Ken: No problem, let's see... ah! The firefighting pipe needs to be fixed! It's raining up here!

Anna: Need a hand up there?

Ken: I'm OK, but the pipe definitely needs replacing! Please give me some duct tapes, a screwdriver, and some spare parts!

Anna: OK! Hang on. I have to get Jason to help us out. He is the plumber!

Ken: Could you also have Leo come over?

Anna: All right! Oh, shoot! Power outage! Cut the power!

(2 hours later)

Ken: Phew. That was a close call, thanks, guys!

Anna: Well, at least we have got it under control. I need to call the supervisor...

Questions

1. How many problems occurred during the day?
2. How did they solve the problems?

Ⅲ Core Knowledge of Solving the Problem

用need to be done来表示某事需要解决，用have someone do something/get someone to do something来寻求他人的帮助（让某人做某事）。have/get something done 代表必须解决的某事（不论任何人）强调某事被做或被完成。	
need to be done = need doing（表被动）	
need to be done	Three engineers need to be hired before Saturday.
need doing	The broken scaffold needs replacing.
get someone to do something	We must get our welder to weld those pipes.
have/get something done	I should have/get the drawings done by Friday!

Ⅳ Exercise and Activity

用have someone do something/get someone to do something或者 have/get something done来解决表格中的问题。(Is it urgent? 哪个比较紧急?)

Put off the fire.		Where can I find our materials?	
The investment delayed.		The walls are not thick enough!	
There is a quality problem.		How much money do we need to spend this month?	
No one drives the crane!		Pipes are leaking.	

Chapter C Suggestions and Advice 建议和意见

Ⅰ Key Words and Expressions

you see 你想想	miscalculation 误算	didn't mean that 不是这个意思	Would it be OK? 可行吗?
budget list 预算单	original 原始的，原创的	difficulties 困难	irresponsible 不负责的
finest 最好的	ensure 保障	work something out 找到解决的方法	25 percent 百分之二十五
operate 操作	It's a good idea to... ……是一个好主意	discount 优惠，打折	or something 类似的
offer 提供	compare 对比	Why don't we...? 为何不……?	vacation 假期

 Ⅱ Read the Material Below and Answer the Questions

Dirac: Hello, Mr. Dawson. This is the cost budget list for the materials and equipment. Please have a look.

Dawson: Well, to be honest, it's quite expensive... and it's more than I expected.

Dirac: Oh, how come? You see Mr. Dawson, our company provides the finest materials and the most excellent workers to operate them, so it seems that the price is quite reasonable.

Dawson: Ah... Let's see, it says here that the total cost is about 1,000,000 dollars, but our original budget was only about 750,000 dollars. So it's almost like a 25% difference. Your rival company CNPC has offered us a deal of 800,000 dollars for the list of materials.

Dirac: Well, I guess there might have been some miscalculations that happened while we were working on the project's original budgets, for this we would like to apologize to you, but we have to ensure the project's quality, right? If you still don't believe us, it's a good idea to compare the materials and equipment of ours with CNPC's.

Dawson: Dirac, I didn't mean that! It's just that we would have some difficulties to make such a change towards the budget. I think you should talk to your boss about it and see if he could work something out.

Dirac: I understand, I would discuss this matter with Mr. Zhang and see if we could give you a discount or something. Would that be OK?

Dawson: Wonderful.

Questions

1. What are the speakers talking about?
2. Did Dirac work out a solution for Mr. Dawson?

 Ⅲ Core Knowledge of Making Suggestions and Giving Opinions

1. Whenever a disagreement happens.

Opinions	Suggestions	Ask questions	Apologize	Solutions
You see...	You should...	How come?	Sorry about that.	We are going to...
I think...	You could...	Why is that?	Truly sorry.	We have done... already
It seems to me...	Why don't you...?	What has happened?	I would like to apologize.	We are doing... right now to...
If you ask me...	It's a good idea to do...	What's the problem?	It was our fault.	Our company is willing to...

2. Learn to react and explain.

Demands	Opinions	Problems	Solutions
I think it's too expensive.	Well in my opinion... It would be a good idea to compare our materials... Why don't we give you a...（妥协）	The project is delayed!	How come? What happened? It seems to me... It's a good idea to... We would like to apologize.
You need to send more men to the site!		John should be fired.	
This material is better than that one.		Adrian was too irresponsible.	
I need a vacation now!		You need to replace the current project manager Ken.	

 Ⅳ Exercise and Activity

解决以下问题。

We spend too much money on the service.	
The material's quality is terrible!	
The EHS supervisor sleeps at work!	
I don't think we should cooperate any more!	

（趁热打铁，课后练习见146页）

Appendix Claims and Compensations, SOI, PQ, Authorization and Rejection Letters 索赔和赔偿，意向函，资格预审函，授权函，拒绝函

Part 1 Claims and Compensations

1. Key words and expressions.

abide by contract 遵照合约要求	claim for construction period 工期索赔	arbitration 仲裁	default party 违约方
refuse the claim 拒绝索赔	claim management 索赔管理	notice of default 通知违约方	potential dispute 替在争议
satisfy the claim 满足索赔	detailed claim 索赔细节	accept claim 同意索赔	claim time period 索赔时间
the act of disaster 灾难原因	final claim 最后索赔要求	waive claim 放弃索赔	claim assessment 索赔评估
the amount of the claim 索赔额度	unreasonable claim 不合理索赔要求	amicable settlement 友善和解	interim claim 中间索赔
赔偿的英文为compensation；索求，索赔可用claim。			

2. Claim procedure for construction works (工程索赔程序).

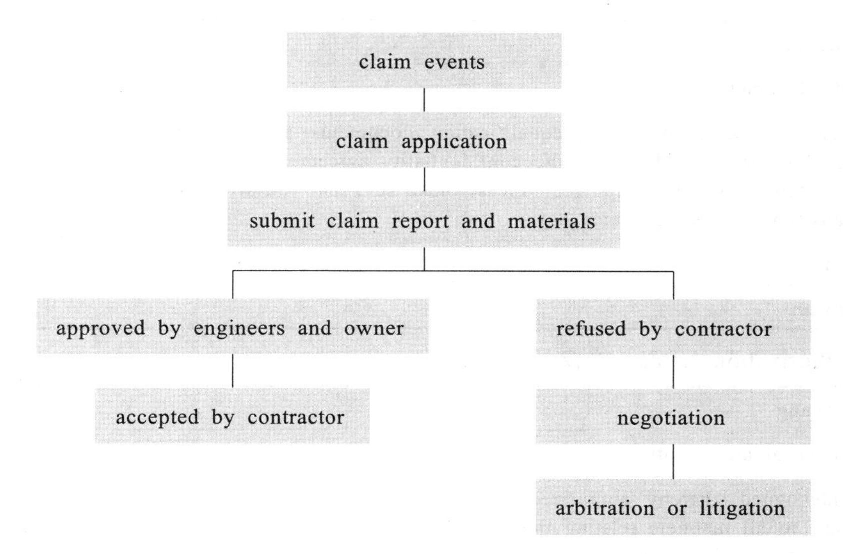

Part 2 SOI, PQ, Authorization and Rejection Letters

在工程竞标过程中,有四种邮件是最常见也是必须学会如何回复的,那就是SOI (scope of interest)意向函,PQ letter资格预审函,authorization letter授权函和rejection letter 拒绝函。而这些邮件则构成了整个竞标过程中最重要的开环和闭环。甲方一般发送这些信函,而乙方则要对这些信函进行回复。

1. SOI letter (意向函).

To: CCOC representatives
Subject: SOI for the KNDP project
We are sincerely congratulate CCOC for qualify to bid for our KDNP phase 3 project,and the attached files are our scope of interest. Please read through the scope carefully, and please send us further information towards your project execution plans and submit a proposal before 20th of July 2019. Thank you! Look forward to hearing from you soon.
Sincerely yours
Marian Timberlake

SOI reply (意向函回函).

To: Ms. Timberlake
Re: SOI for the KDNP project.
On behalf of CCOC, I would like to thank KDNP for this great opportunity to work with you, and CCOC is willing and able to submit a proposal within 3 working days, and other information related to the SOI will be submitted to KDNP before Friday this week. Thank you again for the opportunity and have a good one!
Best wishes
Gabriel Stewart

2. PQ letter（资格预审函）.

To: Mr. Jackson
Subject: PQ documents
The attached documents are our prequalification documents as for your reference, and the hard copy of PQ document together with the confidentiality agreement will sent to you by our company's representative this afternoon at 3:00 p.m. Please make sure that you are in the office for the agreement's arrival. Thank you and have a nice day.
Best regards
Helen Morrison

3. Authorization letter（授权函）.

To: Terry Zhang
Subject: Letter of authorization
We are undersigned, hereby authorized Mr. ________ with ID NO. 4059879198005031233 to act on our behalf in all manners relating the KDNP project, including signing all the documents relating to this matter. Any and all acts carried out by Mr. ________ on our behalf should have the same effect as acts of our own. The authorization is valid until further notice from 1st of August 2019. Thank you for reading, and congratulations.
Best regards
Jay Hoffman

Authorization reply（授权函回函）.

To: Mr. Hoffman
Re: Letter of authorization
I'm writing this letter on behalf of CCOC to thank you for giving me your trust and the opportunity to work for you, and I have also attached further information towards the project for your reference. Although this is the first time we cooperate together, yet I believe we can guarantee you a job well done. Thank you for the letter, and good luck.
Yours truly
Terry Zhang

4. Rejection letter（拒绝函）.

To: KDNP
Subject: SOI for the KDNP EPC project
CCOC would like to first thank you for the invitation to participate in the bidding of this project, we are deeply appreciated your trust towards us. However, while considering the limited bidding time, our proposal, our workload, and our focus on the existing bidding as well as on-going projects, we have to decline this wonderful opportunity. Once again, we deeply appreciate and thankful for your trust on us and we are all look forward for more future cooperation with KDNP!
Best regards
Richard Du

Answers

Chapter A Clarification Emails

Ⅱ Read the Material Below and Answer the Questions

1. “Flooded” here means the whole floor was filled with water.
2. They are impressed with his experience in the field of lubricant oil.
3. He needs an explanation, and a solution to the delay.

Chapter B Helping Hand

Ⅱ Read the Material Below and Answer the Questions

1. Three incidents have happened on that day.
2. They have the power cut down, get the pipes duct taped, and have the pipes replaced.

Ⅳ Exercise and Activity

用have someone do something/get someone to do something或者have/get something done来解决表格中的问题。

Put off the fire.	We need to have the firefighter put off the fire!	Where can I find our materials?	I would get it done by calling the suppliers.
The investment is delayed.	We need to get the owner's representatives to pay off the cost.	The walls are not thick enough!	I'm going to get it checked and examined.
There is a quality problem.	I'm going to call the quality inspector to check the problem.	How much money do we need to spend this month?	Richard should have this checked before Friday.
No one drives the crane!	Get the crane operator to call the driver!	Pipes are leaking.	We should have Shirley fix this problem!

Chapter C Suggestions and Advice

Ⅱ Read the Material Below and Answer the Questions

1. They are talking about the price of the equipment.
2. Technically he didn't, but he promised to talk to his boss and work something out.

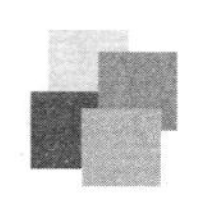

Ⅳ Exercise and Activity

解决以下问题。

We spend too much money on the service.	It seems to me that our company provides the finest workers to operate them, so it's totally reasonable.
The material's quality is terrible!	Well I'm truly sorry about it. I'm going to send our best engineers to fix it for you! Why don't we sit down and talk through this matter?
The EHS supervisor sleeps at work!	Sorry to hear that, but it seems that he has been working over time for the past few weeks, so I hope that you could understand his exhaustion.
I don't think we should cooperate any more!	Oh how come? It seems that we have such a good cooperation together! We should talk about it.

Class 6

Business Changes
行业变化

Chapter A Reporting News 报告新闻

I Key Words and Expressions

monthly magazine 月刊	national park 国家公园	team building 团建	compensate 赔偿
interesting 有趣的	executive's office 总裁办	serious accident 严重事故	steel bar 钢制栏杆
board meeting 董事会会议	sounds good/fun... 听起来不错	severe storm 严重的风暴	institute of architecture 建筑机构
marathon 马拉松	enrollment 签到	temporary office 临时办公室	save a spot 留位置
investigate 调查(事后)	sign up 报名	KPI 关键业绩指标	regulations 规章制度

II Read the Material Below and Answer the Questions

Henry: So Kelly, what are you reading?

Kelly: Oh, I'm reading our company's monthly magazine.

Henry: See anything interesting?

Kelly: Em... Let's see, the 4th board meeting was held on the 5th of July.

Henry: Oh, listen to this: a marathon was organized in the national park last Tuesday by the executive office.

Kelly: That sounds fun, pity that I didn't go!

Henry: The annual BIM training program was finished this Monday, and 50 employees were trained. Ah... what else? The English training enrollment was closed because too many students signed up for it!

Kelly: What! Really! I should call Richard to save a spot for me!

Henry: Interesting! Oh look, here is something else we could go for. There are going to be a few team building activities held on Sunday this week. I think they will go boating and jogging.

Kelly: Great! Oh! Look at this one: a severe storm caused a serious accident in Fujian!

Henry: Really? What happened!

Kelly: It says that a temporary office was crushed by a falling steel bar. Emergency engineers were gathered to solve this problem. Five people were injured.

Questions

1. What are the activities that happened this month?
2. What happened to the site in Fujian?

Ⅲ Core Knowledge of Simple Past Passive

Active Speech 主动语态		Passive Speech 被动语态	
Focus on the doer. 关注主语		Focus on the receiver and the result. 关注宾语	
Event	Doer	Event	Receiver
John created the new KPI system.	John	The new KPI system was created.	the new KPI system
Ben held a meeting to increase workers' efficiency.	Ben	A meeting was started to increase workers' efficiency.	a meeting
BOE increased their budget last year.	BOE	The budget was increased by BOE last year.	the budget
西方人喜欢把注意力放在结果上,所以很多正式场合或新闻和报告中都会使用到被动语态。被动语态中,"谁做的"并不重要,关键在于"做了什么,什么受到了影响"。			

Tip: 如果主动语态的主语是个对事件十分重要的元素,则可以在被动语态的句后加上by+doer来强调行为人和物。一般情况下, 谁(who)做的并不重要。

Active (do, does, did)	Passive (be done)	
I broke the regulations.	Regulations were broken by me.	动词过去式变成were+过去分词,主语I 变成宾语me。

Ⅳ Exercise and Activity

将以下的主动语态变成被动语态。

Dickens fired Tom this morning.	Tom was fired by Dickens this morning.
Our board manager Mr. Lin did a presentation to talk about his life last Friday.	
Doctor Mann opened a new institute of architecture in 2011.	
Our boss told us to work overtime this weekend.	
BOT cancelled their contract with us this Thursday.	
NPO sent 5 experts to investigate our accidents on the site.	
Our company paid 480,000 dollars to compensate NPO's loss in January.	

Chapter B Business Changes 行业变化

I Key Words and Expressions

shortage 短缺	outsourced 外包	new energy 新能源	renovate 翻新
language expertise 语言精通	comfort zone 舒适区	financial support 经济支持	manufacture 制造
salary 薪酬	compete 竞争	bright future 美好前景	knock down 推倒
laid off 失业	encourage 鼓励	be forced to 被迫	hall 大堂
not in a good shape 糟糕	explore 探索	cut down on the cost 削减开支	pay attention to 集中注意力

II Read the Material Below and Answer the Questions

Sally Toyoshima（Tokyo）

In recent years, our business has had a shortage of employees with professional skills and language expertise. So a lot of skilled employees are being recruited from places like Singapore, Malaysia, and Indonesia, and they have been paid with a really high salary.

Hans Morgan（Belgium）

Many young people nowadays have been laid off because our country's economy is not in a good shape, and also a lot of opportunities are being outsourced to workers overseas. It has become a huge problem for our society, because even though things like this happened, a lot of young people still want to stay in their comfort zone and are not willing to compete as much as they need to.

Morris Harrison（London）

Well, many big companies today are being encouraged to explore into new energy and high-tech fields. I think it's a great idea, and great financial support has been given by our government. So I believe that we could have a bright future in this area.

Sara Catherine（San Francisco）

Our company has a huge problem with our working efficiency. Although a lot of work has been done, and much training has been made, the problem with efficiency still needs attention.

Curtis Ming（Shanghai）

Many companies nowadays focus on the price instead of the quality during a bid. So, many construction companies have been forced to cut their budget down, which has caused many quality incidents. Many efforts have been made by our company to cope with this problem, but things haven't changed much.

Questions

1. Why are many young people laid off?
2. What do today's companies mainly focus on?(请用被动语态回答)

Ⅲ Core Knowledge of Past Achievements and Recent Incidents

Active Speech (be doing) 进行时主动语态	Passive Speech (be being done) 进行时被动语态
Many companies are outsourcing jobs to other countries.	Jobs are being outsourced to other countries.
Our boss is working on a new project.	A new project is being worked on.
CNPC is decreasing its revenue to find more opportunities.	Revenue is being decreased by CNPC.

Tip:Remember, in passive speech, the "doer" is not important.

Ⅳ Exercise and Activity

1. 重写以下句子,用方格中的词语开头。

Many poor countries are sending really cheap employees to our manufacture industries.	cheap employees
They have completely renovated the building; they have built up 6 new offices and 2 new halls by knocking down many walls on the first floor.	the building
They are promoting their newest equipment to many companies.	their newest equipment

2. 填空,完善句子。并想想,在你们的公司也会这样吗?

More females ____________ encouraged to join the engineering work force.
EHS trainings ____________ organized as a compulsory course in many construction companies.
Many workers in our company ____________ told to learn English.
Our salary ____________ raised by our project manager these few months.

Chapter C Installation Instructions 安装要求

 I Key Words and Expression

FFU(fan filter unit) 空气过滤器	master 掌握	final drawing 最终稿	superior 上层领导
various 多种多样	setting out basis 陈述原理/确定原则	crude oil tank 原油桶	punish 惩戒
checked up 检测	assembling machine parts 组装机械部件	put into use 启用	instruction/operation menu 说明书
fitter 钳工	adjustment 调整	as well as... 和……一样	assembling machine 组装器械
mount machines 安装机械	in accordance to 遵照	approved 批准	Write down any other words you don't know and ask!

 II Read the Material Below and Answer the Questions

Phil: The FFU were installed, good.

Patrick: Well yes sir, I'm Patrick, and the installation of various kinds of equipment is a part of my job.

Phil: Could I start the system now?

Patrick: Sir you may not, because the installation has just been finished, and in order to guarantee its safety, it should be checked up a couple of times before we put it into use!

Phil: Ah, all right, that seems like a good idea. So, what do you mainly do here?

Patrick: I'm a fitter. I mainly mount machines and equipment such as factory lathes (车床), cranes (起重机), and things like that.

Phil: Sounds good, so how many techniques have you mastered?

Patrick: Let's see, for example, installing the equipment, cleaning and assembling machine parts should be mastered. Oh! Equipment adjustment and testing must be learned as well as the others.

Phil: Phew. That's many things. Were all these equipment installed and mounted by your team?

Patrick: They were! They were all installed in accordance with the final drawing.

Phil: What can I say, job well done Patrick!

Questions

1. What's the relationship between the two speakers?
2. What does Patrick have to learn before he enters the job?

Ⅲ Core Knowledge of Active and Passive Speech of Modal Verb

1. 情态动词主动语态和被动语态对比。

主动语态		被动语态	
should do	may do	should be done	may be done
would do	might do	would be done	might be done
could do	shouldn't do	could be done	shouldn't be done
must do	couldn't do	must be done	couldn't be done
情态动词的被动语态一般用在某个应当被做的事件上,谁去做不重要,重要的是这件事的执行。而主动语态则强调动作的执行者。比如说You should finish your homework. 关键是You。而Homework should be finished. 则是以完成作业这个行为优先,强调某事被做、被完成。			

Ⅳ Exercise and Activity

1. 把以下主动句变为被动句。

You should learn how to manage a team.	How to manage a team should be learned.
We must punish Joseph for smoking near the crude oil tank!	
Our project manager should stop the noises coming from the site at night!	
They might approve the proposal by the end of next week.	
They could replace the procurement manager easily.	
They should warn our engineers to wear PPE before entering the site.	

2. 讨论以下观点,然后模拟做出决定,或者创造一些应当执行的决定。

We should get a raise right now!	Absolutely! Because I have been working my heads off for this job!
Our society should encourage more young people to start their own business!	

(趁热打铁,课后练习见146页)

Appendix Construction and Engineering News 建筑工程新闻

Reading Construction News Reports

UK Construction in Recession After Largest Output Fall in Five Years

Construction has slid into recession after the largest quarterly fall in output for five years, according to the Office for National Statistics.

Output in construction fell by 0.7 percent in the last three months to September, which represented the largest quarterly fall since the third quarter of 2012 and follows a 0.5 percent decline in the second quarter of 2017. Two quarters of falling output puts the industry in a technical recession. However, output still remains above the pre-financial crisis peak. The figures are based on forecasts and responses to the ONS business survey for September. Earlier this month the Market / CIPS purchasing managers' index, which measures construction activity, posted its first contraction in 13 months after warning that Brexit (脱欧公投) uncertainty was slowing new projects.

CASE Introduces New Minimum-Swing Radius Excavator

CASE Construction Equipment has introduced the CX245D SR minimum-swing radius excavator to its D Series line-up, designed for confined work areas. The CX245D SR features a compact counterweight and modified boom placement which results in a productive and maneuverable excavator that is suited for restricted conditions, such as road and bridge work, residential projects and urban construction.

Questions

1. Why is UK's construction industry facing a recession?
2. What environment is the new excavator suited for?

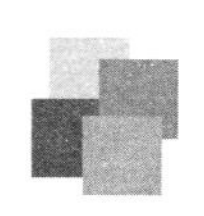

Answers

Chapter A Reporting News

Ⅱ Read the Material Below and Answer the Questions

1. A board meeting, a marathon, BIM and English training program, and some team building activities were held by the company.

2. A serious accident was caused by a severe storm.

Ⅳ Exercise and Activity

将以下的主动语态变成被动语态。

Dickens fired Tom this morning.	Tom was fired by Dickews this morning.
Our board manager Mr. Lin did a presentation to talk about his life last Friday.	A presentation was done by our board manager Mr. Lin.
Doctor Mann opened a new institute of architecture in 2011.	A new architecture institute was opened by Doctor Mann.
Our boss told us to work overtime this weekend.	We were told to work overtime this weekend.
BOT cancelled their contract with us this Thursday.	Our contract was cancelled by BOT.
NPO sent 5 experts to investigate our accidents on the site.	5 experts were sent to investigate our incidents.
Our company paid 480,000 dollars to compensate NPO's loss in January.	480,000 dollars was paid to compensate NPO's loss in January.

Chapter B Business Changes

Ⅱ Read the Material Below and Answer the Questions

1. A lot of employees are being recruited from places like Singapore.

2. The price instead of the quality is focused on by many companies.

Ⅳ Exercise and Activity

1. 重写以下句子，用方格中的词语开头。

Many poor countries are sending really cheap employees to our manufacture industries.	cheap employees
Cheap employees are being sent to our manufacture industries by many poor countries.	
They have completely renovated the building; they have built up 6 new offices and 2 new halls by knocking down many walls on the first floor.	the building
The building has been completely renovated...	
They are promoting their newest equipment to many companies.	new equipment
Their newest equipment is being promoted...	

2. 填空，完善句子。并想想，在你们的公司也会这样吗？

More female are being encouraged to join the engineering work force.
EHS trainings have been organized as a compulsory course in many construction companies.
Many workers in our company are being told to learn English.
Our salary has been raised by our project manager these few months.

Chapter C　Installation Instructions

Ⅱ Read the Material Below and Answer the Questions

1. Their relationship might be a fitter and his superior.

2. Setting out the basis, installing the equipment, cleaning and assembling machine parts and things like that should be mastered. Equipment adjustment and testing must be learned as well as the others.

Ⅳ Exercise and Activity

1. 把以下主动句变为被动句。

You should learn how to manage a team.	How to manage a team should be learned.
We must punish Joseph for smoking near the crude oil tank!	Joseph should be punished for smoking near the crude oil tank!
Our project manager should stop the noises coming from the site at night!	Noises should be stopped!
They might approve the proposal by the end of next week.	Proposal might be approved by the end of next week.
They could replace the procurement manager easily.	The procurement managers could be replaced easily.
They should warn our engineers to wear PPE before entering the site.	Engineers should be warned to wear PPE before entering the site.

2. 讨论以下观点，然后做出模拟决定，或者创造一些应当执行的决定。

We should get a raise right now!	Absolutely! Because I have been working my heads off for this job! The salary should definitely be raised.
Our society should encourage more young people to start their own business!	Definitely. Young people means energetic blood and creativity, and therefore I totally agree with you!

Class 7

What Would You Do?

你会怎么做?

Chapter A Imagining 设想

I Key Words and Expressions

grab a bite 吃点东西	construction firm 建筑公司	PVC thickness PVC厚度	major incident 重大事件
examining 检测	day dream 白日梦	working hours 工时	a large sum of money 一大笔钱
check list 检测表	realistic 现实的	excited 兴奋的	current condition 当下状态
far away 遥远	help someone out 帮助某人	promotion 升职	career 职业生涯
economically independent 经济独立	in debt 欠债	foreign 外国的	wealth 财富

Ⅱ Read the Material Below and Answer the Questions

Frank: Lawson, it's 5:30, you want to grab a bite and go home?

Lawson: I don't know. I still have so much examining to do today. This checklist needs to be finished before tomorrow.

Frank: Wow, that's a lot of work! Why are you working so hard recently?

Lawson: I just got married, and my wife and I are renting a really small apartment. If I could make more money, I would definitely buy our own place to live.

Frank: Well, I think you would also spend less time with your family if you keep working like this.

Lawson: I know and I wish I didn't work so far away from home, then I could have more rest each day.

Frank: Yeah, I guess we have the same problem. I just don't have enough money! I wish I could be more economically independent, and if I were rich, I would travel around the world, buy a new car or open my own construction firm.

Lawson: Haha, that sounds like a day dream! If I weren't so busy, I would go sign up for a technician class! I have always wanted to be a technician!

Frank: Oh well, realistically, I want to help my parents out financially, because they have been in some debt for the past few years, and they need a large sum of money to pay off!

Lawson: Good luck with that! Well, I guess I need to test the PVC thickness now!

Frank: All right, let me help you out!

Questions

1. What are the speakers' current conditions right now?
2. What's Frank's real problem?

Ⅲ Core Knowledge of Talking About Imaginary Things

假设	时态变为过去时	行为
I wish	I could do...	I would do
If	I were	I could do
I hope	I did	I will do
假设一个完成了的状态,用would, could, will 来描述假设中的行为(不能或者难以实现的行为)。		

Tip:虚拟语气一定是现实的相反面。

Dreams (If/I wish/I hope + simple past)	Action (would/could+simple present)
I wish I were rich.	I would buy myself a Benz.
If I could live closer to my parents.	I could spend more time with them.
I hope our owners were more easygoing.	Then I could have easier working hours.
If our new recruits were hard-working.	I would teach them more skills.

Tip:虚拟语气中的行为也是假设的行为。

Ⅳ Exercise and Activity

给以下假设加上行为或给行为加上假设。

If I were the boss of this company,			I would make more friends.
I wish I could be a foreigner,			I could play basketball all day.
If I could change my work,			I would be really excited.
I wish I weren't so busy,			I could get a promotion

Chapter B Regrets and Summarizing 反省与总结

 I Key Words and Expressions

disputes 争议	yell 吼叫	lost track 忘了	overslept 睡过了
polite 礼貌的	cursed 咒骂	hung up 挂机	casual inspection 突击检查
normally 一般来说	lost temper 情绪失控	brandy 白兰地	corrosive chemical 腐蚀性化学品
exact disagreements 具体争议	center of experts 专家中心	first thing in the morning 早上第一件事	disapprove 不通过
drunk 醉酒	switch off 关机	patient 耐心的	technical report 技术报告

 II Read the Material Below and Answer the Questions

Jessica's View

Last night, my owner called me to discuss the disputes towards our newest contract and samples. Usually, I would have been really polite, and I would have just told him the exact disagreements on the contract and got it over with, but it was already 1 a.m. in the morning, and he was drunk. Finally, he yelled at me, so I got furious and cursed him for at least 5 minutes and hung up on him! Gosh, I felt awful after I hung up and I called him back, but no one answered! I really shouldn't have lost my temper! I should have just told him that I could meet him on the next morning in the center of experts! Also, I should have just switched off the phone at night! Now my job is at risk!

Bryant's View

Well, I was drunk last night and lost my track of time. So I called Jessica, PCP's project manager to discuss the disputes of our contract and samples. I could have just stopped the conversation when I realized that she was mad, and I would have just said sorry to her before she hung up and shouldn't have yelled at her like that. However, I was too drunk to do so! I shouldn't have joined that stupid party my friend organized, and I shouldn't have drunk so much brandy! She must have been furious! I really need to apologize to her as the first thing in the morning!

Questions

1. What things should Jessica have done to avoid this incident?
2. What things shouldn't Bryant have done?

Ⅲ Core Knowledge About Hypothetical Past

<table>
<tr><th>Real Behavior</th><th>Hypothetical Behavior</th></tr>
<tr><td>I have yelled at my boss.</td><td>I should have been more patient...</td></tr>
<tr><td>I overslept and missed an important meeting.</td><td>I should have switched the alarm on!</td></tr>
<tr><td>Josh was playing his cell phone when his supervisor came.</td><td>He could have just looked around before he did it.</td></tr>
<tr><td>Michael forgot to bring his wallet and phone out when he was having a dinner with his clients.</td><td>He could have just asked his colleagues to help him out! Instead, he ran away!</td></tr>
<tr><td colspan="2">一般should/could/would have done 都用在不可逆的事实发生之后,表示对自己行为的反思、责备或者后悔,并提出别的可能性。通常用于吸取教训。</td></tr>
<tr><td colspan="2">注意:should have done something 意为应当做但是没做, shouldn't have done something 意为不应该做但是做了。</td></tr>
</table>

Ⅳ Exercise and Activity

1. 思考标黑的意外事件,如果是你,你会怎么检讨?(给出至少两个假设)

<table>
<tr><td>Our suppliers failed to deliver a really important piece of equipment to us, which caused a serious delay in our project.</td></tr>
<tr><td>He should have just double checked everything before he delivered them!</td></tr>
<tr><td>The local safety control department organized a casual inspection yesterday to one of our projects, but no EHS staffs was on the site that day, and a lot of the project crews weren't wearing any PPE (安全设备) equipment during the inspection.</td></tr>
<tr><td></td></tr>
<tr><td>Harris saw some people were smoking in a car near the corrosive chemical containers and threw out the cigarette butts (烟蒂) and some garbage from the window. So he picked up that garbage and threw it back into the car.</td></tr>
<tr><td></td></tr>
<tr><td>The owner from Intel has disapproved our procurement proposal and had a huge argument with our managers. Now the project is on hold, and the loss has become larger.</td></tr>
<tr><td></td></tr>
<tr><td>Tim's company failed to deliver the drawing and technical clarification report, as a result a serious incident happened. Now Tim's company is facing a large sum of compensation.</td></tr>
<tr><td></td></tr>
</table>

2. Tell your partners some mistakes you have made in your life and try to correct them.

Chapter C Speculating 猜测

 Ⅰ Key Words and Expressions

cement 水泥	appointment 预约	bull dozer 推土机
see someone anywhere 看到某人了吗	cut off 切断	roller driver 压路机司机
transfer line 转接	be tied up with 被束缚	ram pile driver 打钻机司机
Don't hang up. 别挂断。	examining center 检测中心	road paver 铺路机
Be at your service. 为您效劳。	try one's luck 碰运气	non-signal area 无信号区域

 Ⅱ Read the Material Below and Answer the Questions

Richard: Hello this is CCOC. I'm Richard. What can I help you?

Kate: Hey Richard! Did you see Sara anywhere? I really need to find her before I'm doing the test on our cement.

Richard: Have you told her about this before? She might have forgotten. Let me transfer you to line 3, the pharmaceutical department. She might be there. Just don't hang up.

Kate: OK! Thanks a lot Richard! Catch up with you soon!

Sara: Good afternoon, this is the CCOC pharmaceutical department. How can I be at your service?

Kate: Yes, I'm Kate Underwood, from the quality control department. I have an appointment with Sara Mathews. Is she there?

(cut off)

Sara: Hello, who is the person you look for? Hello?

Kate: Oh no... the connection must have been cut off! (Phone ringing) Oh hi, Kate Underwood speaking!

Sara: Hey Katie! Where are you? I've got tied up with my work! Hello? I can't hear you! She could have been in some non-signal area!

Kate: Finally! Sara! You need to come over for the testing. We are late.

Sara: Oh no! It's 4:30 already! The examining center must have been closed by now.

Kate: Well, let's try our luck and check it out first.

Questions

1. Why was Kate looking for Sara?
2. Why was Kate's signal so bad? What might have happened?

Ⅲ Core Knowledge of Speculation

1. Speculating.

He might be a ram pile driver.

He may be a bull dozer driver.

He must be a roller driver or a paver.

He could have been in a hurry to pave the roads.

He must have been busy in the project.

He might have forgotten to attend the safety training this morning.

Modal Verbs	Speculating a State	Speculating a Situation or Action
must (100% sure)	be	must + have done
could (80 ~ 90% sure)	be	could + have done
might (less than 50%)	be	might + have done
may (less than 50%)	be	may + have done
must/could/can/may/might have (been) done一般被用来描述对已完成行为的猜测。		

2. Making phone conversations.

Formal Phone Call Expressions			
Hello this is... speaking.	This is the... department.	Please don't hang up.	I'm calling for...
How may I be at your service?	Let me transfer you to line 3.	Call you back later.	Please dial this number.
Hi, how can I help you?	Oh I've got cut off!	Hold on. Let me write it down.	Sorry, please say that again.

Ⅳ Exercise and Activity

1. 阅读以下情况，猜测发生了什么事。

Our appointment with the owner was 3 hours late. He was nowhere to be found!
The current location of the meeting has to be changed!
Our owner's investment was delayed.
Our drawings were sent back by the design bureau.

Continued

The headquarter of our company has been moved to Thailand.

2. 与同桌进行一次模拟的电话对话。

题目	回答
Hello sir, Jeremy's metal goods!	
This is SIEMENS. What can I do for you?	
Customer services, what can I do for you?	

(趁热打铁,课后练习见147页)

Appendix Items for Quality Examination 常用建材检测项目

Material Name 材料名称	Examining Items 检测项目
cement 水泥	normal consistency, initial bond, final bond, soundness, mortar strength, fineness
sand 砂	sieving, mud content, clod content, sulfide, light components, chloride content, mica content, organic matter content, apparent density, bulk density, consistency
stone 碎石	stone crushing index value, needle like flaky particle content, density, sieving, mud content, clod content, consistency, void ratio
water 水	PH value, chloride content, sulfate content
admixtures 外加剂	solid content, density, water reduction rate, bleeding rate, setting time differences, compression strength rate, chloride content
reinforced steel 强化钢筋	ration of elongation, yield strength, ultimate tensile strength, bending test
wood 木材	water content
brick 砖	bending strength, compression strength, anti-freezing performance
clay and cement tile 黏土,水泥瓦	anti-bending load, water absorbing capacity
plastic drain board 塑料排水板	longitudinal drainage, filter film infiltration parameter, filter film equivalent aperture, filter film tensile point, longitudinal moisture-free lateral wet process
geotextile 土工布	unit area, thickness, effective pore size, ultimate tensile strength, elongation percentage, CBR picture strength, perpendicular infiltration parameter

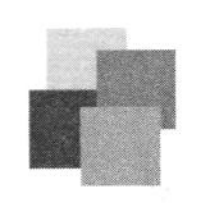

Continued

Material Name 材料名称	Examining Items 检测项目
PVC civil film 聚氯乙烯土工膜	unit area, film thickness, breaking/rupture strength, elongation at rupture, tearing strength, CBR puncture strength, anti-infiltration strength, infiltration parameter, corrosion resistance
geogrid 土木格栅	ultimate tensile strength, ration of elongation, tensile strength
concrete sample 混凝土试件	compression strength
masonry mortar 砌筑砂浆	fluidity, compression strength
petroleum asphalt 石油沥青	penetration, extensibility, softening point
bitumen waterproof roll 沥青防水卷材	impermeability, heat resistance, water absorbency, ultimate tensile strength, flexibility
bituminous adhesive 沥青胶	heat resistance, flexibility, cohesive force

Answers

Chapter A Imagining

Ⅱ Read the Material Below and Answer the Questions

1. Lawson needs a new apartment. Frank needs to pay off his parent's debt.
2. Frank needs to pay off his parent's debt.

Ⅳ Exercise and Activity

给以下假设加上行为或给行为加上假设。

If I were the boss of this company,	I would give my staffs a long vacation	If I were famous enough,	I would make more friends.
I wish I could be a foreigner,	then I could speak excellent English.	If I weren't working so hard,	I could play basketball all day long.
If I could change my work,	then I would be an artist.	If I could get a raise,	I would be really excited.
I wish I weren't so busy,	I would spend more time with my family!	If I weren't so lazy,	I could get a promotion.

Chapter B Regrets and Summarizing

Ⅱ Read the Material Below and Answer the Questions

1. She really shouldn't have lost her temper. She should have just told him that she could meet him on the next morning in the center of experts. And she should have just switched off the phone at night.

2. He shouldn't have joined that party, shouldn't have called her, and shouldn't

have been drunk and yelled at Jessica.

Ⅳ Exercise and Activity

1. 思考标黑的意外事件,如果是你,你会如何检讨?(给出至少两个假设)

Our suppliers failed to deliver a really important piece of equipment to us, which caused a serious delay in our project.
He should just double checked everything before he delivered them!
The local safety control department organized a casual inspection yesterday to one of our projects, but no EHS staffs was on the site that day, and a lot of the project crews weren't wearing any PPE equipment during the inspection.
Our project managers should have called the EHS staffs on the site earlier to tell them about the inspection, who should have just obeyed the safety rules, and have wore the PPE equipment.
Harris saw some people were smoking in a car near the corrosive chemical containers and threw out the cigarette butts and some garbage from the window. So he picked up the garbage and threw it back into the car.
He should have just called the security instead of doing such an aggressive action.
The owner from Intel has disapproved our procurement proposal and had a huge argument with our managers. Now the project is on hold, and the loss has become larger.
We should have been more patient, and could have just arranged another meeting with the owner.
Tim's company failed to deliver the drawing and technical clarification report, as a result a serious incident happened. Now Tim's company is facing a large sum of compensation.
Tim's company should have discussed with the clients to stretch the deadline a little.

2. Open.

Chapter C Speculating

Ⅱ Read the Material Below and Answer the Questions

1. Kate was looking for Sara to make an examination towards the quality.
2. Her signal might have been cut off. She could have been in a non-signal area.

Ⅳ Exercise and Activity

1. 阅读以下情况,猜测发生了什么事。

Our appointment with the owner was 3 hours late. He was nowhere to be found!
He might have been caught up with some emergency situations.
The current location of the meeting has to be changed!
Our owners could have been dissatisfied with the location we provided.
Our owner's investment was delayed.
They must have forgotten about it.
Our drawings were sent back by the design bureau.
There must have something wrong with the design.
The headquarter of our company has been moved to Thailand.
Our boss might have found a cheaper place to run his business.

2. Open.

Class 8

Socializing
社交

Chapter A Eating in a Project 吃在项目

 Ⅰ Key Words and Expressions

canteen 食堂	a loaf of 一片(面包)	a carton of 一纸盒	spaghetti 意面
project cafeteria 项目食堂	a piece of 一块	steak 牛排	vegetable 蔬菜
starving 饥饿	very little 很少	french fries 薯条	soup 汤
a bottle of 一瓶	not much 不多	sushi 寿司	water purification 水净化
a can of soda 一罐苏打饮料	a few 一点	salad 沙拉	pollution/contamination 污染
takeaways 外卖	menu 菜单	beverage 饮品	roast fish 烤鱼
demolition 拆除	rebuild 重建	loading area 卸货区	harm 伤害

 Ⅱ Read the Material Below and Answer the Questions

Derek: It seems that the new cafeteria has been built. Finally, we could eat something other than takeaways.

Adam: It's good that they built this temporary canteen and office. It saves us much time at work.

Derek: Let's go and see what they have on the menu.

Adam: Oh, not bad. They have burgers, French fries, salads, and all kinds of beverages! Wow look, they also serve roast fish and chicken!

Derek: I'm having a can of soda, some vanilla mushroom soup, and a hamburger to go with it. Ah! They have steaks. Do you want a steak Adam?

Adam: Absolutely! I'm starving! Also some salads and a carton of milk as well, please.

Derek: You like Japanese food? They have sushi here.

Adam: No, thank you. Why don't you get the food, and I will find a seat.

(a while later)

Derek: Here you go. I've also taken a few loaves of bread for you.

Adam: Thank you! So, how is the water

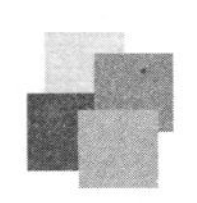

purification work going in your sector?

Derek: It's not going so well. The water from the river was vastly contaminated, and the production line of the old factories has caused a lot of air pollution for the past 22 years!

Adam: Well, I guess a lot of work should be done before we rebuild this old site, but the good news is that the demolition of the old factories is going to be on Monday next week.

Derek: Oh it's 1:00 p.m. already. I'd better go to the loading area to retrieve our equipment.

Adam: Sure, I will go with you. Emm, this soup tastes delicious with the bread!

Questions

1. What did the speakers think of the food in the canteen?
2. What harm did the old factory do to the environment?

C. Some Tips About Food and Lunch Time Conversation

1. Variations of cooking (烹饪手法).

	Fry	Stir Fry	Roast/Bake	Stew	Steam	Raw	Boil
Fish	☺	☺	☺	☺	☺	☺	☺
Chicken							
Lamb/Mutton							
Beef							
Pork							
Vegetables							
Noodles/Rice							
你还能想出其他的组合吗？请写在空白处或者在表格上画笑脸。							

2. Containers and quantities（容器和量）.

a can of	soda	a loaf of	bread
a bottle of	wine	a slice of	cheese
a jar of	sauce	a bag of	chips
a carton of	juice	a box of	chocolate
a liter of/a quart of	milk	可以用 few 和 a few 来修饰可数名词，如 cheese，steaks 等。 a little 和 less 可以用来修饰不可数名词。	
a kilo of/a pound of	beef		

Ⅳ Exercise and Activity

1. What food are you going to put in your home fridge? Fill in the fridge below.

2. Make a lunch menu for your own restaurant. Also tell your partners why you choose these food.

______

______

______

______

______

______

The Menu for ______ Restaurant

So, what do you recommend us to eat?

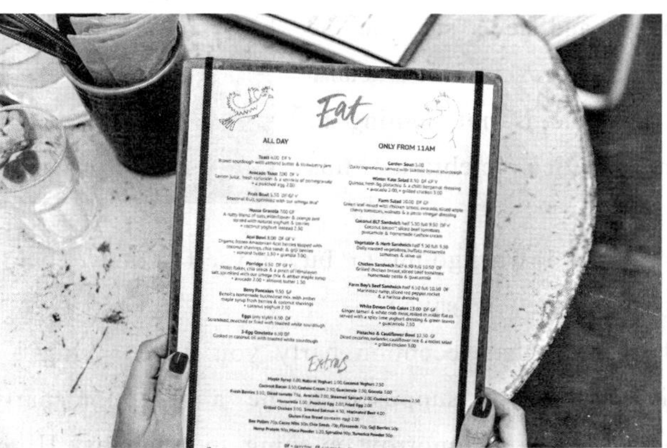

Fried Chicken

______

______

______

______

______

______

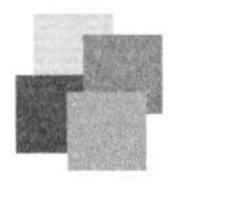

Chapter B Socializing 社交

I Key Words and Expressions

am/is/are supposed to 应当	reserve 预订	bonus 奖金
was/were supposed to 本应	expo 展会	cocktail dress 晚礼服
was going to 本会	stuck in a meeting 被会议拖延	call in the last minute 最后一分钟做改动
reception party 酒会	government official 政府官员	process engineer 工艺工程师
prepare 准备	scientist 科学家	process 工艺

Ⅱ Read the Material Below and Answer the Questions

Susie: Hey Betty, still working?

Betty: Yea, I need to prepare some materials for the meetings tomorrow, and there is also a reception party coming up tomorrow evening. So I need to call the hotel to reserve a hall.

Susie: So what's this meeting about? Where is it?

Betty: It is supposed to be a contract appointment for the P&G project, and it's supposed to be in the Shanghai center tomorrow.

Susie: Isn't the meeting supposed to be in the Beijing center?

Betty: It was supposed to be in Beijing, but Tony has to attend an important Expo today, and he is stuck in the meeting with the government officials.

Susie: Oh dear, so he is not coming to our annual office celebration this year?

Betty: I'm afraid so. I was going to buy myself a new cocktail dress for it, but I guess the money is saved.

Susie: And what about that reception party you were saying?

Betty: Oh, right, there is supposed to be a reception party for scientists and construction experts tomorrow evening in Shanghai Hilton Hotel.

Susie: Hey come on. Let's take a break tonight. There is supposed to be an excellent restaurant downstairs. Let's go and have dinner there.

Questions

1. What is supposed to happen tomorrow?
2. What was Betty's plan?

Ⅲ Core Knowledge of Am/Is/Are Supposed to and Was/Were Supposed to

am/is/are supposed to	
应做没有做	You are supposed to stay in the project. (But you are not.)
应该做的事(与have to 和should相同意义)	I'm supposed to reserve a hall for the reception party.
别人说,听说	It's supposed to end at 10:00 p.m. It's a meeting for government officials.
was/were supposed to, was/were going to	
理应发生但是没发生	It was supposed to rain today!(But it's sunny today.)
was/were going to 也表示原来计划会发生但到了预定时间还没发生,同时也有"曾经想实现"的意思。	Our boss was going to give us a big bonus this year!(But he didn't.)

Tip: It somehow sounds like that you "heard" something is going to happen.

Ⅳ Exercise and Activity

1. 用am/is/are supposed to 或 was/were supposed to 来完成以下句子。

There ________ a school interview ________ be here this Sunday.
Our materials ____________ deliver now.
Our facility management team ________ come next Monday.
The technicians from SHELL ________ send us copies of instructions and ________ organize an open discussion towards the process issues of our project, but they called in the last minute and told us that they couldn't come!

2. 用am/is/are supposed to 和 was/were supposed to 来回答以下问题。

Is there any work you were supposed to do last week but you didn't?
Are there any business events coming up in the next few weeks?
Are you supposed to attend any important meetings?
Who else are you supposed to see today in particular?
Where were you supposed to go if you weren't in this company?
Are there anything interesting happening in our company?

Chapter C Congratulations 恭贺

Ⅰ Key Words and Expressions

invite 邀请	be always doing something 总是坚持做某事	nominate 提名	Seriously? 真的吗?
ceremony 宴会	patent 专利	Noble Price 诺贝尔奖	Good on you! 真棒!
challenging 有挑战的	insulin 胰岛素	invent 发明	proud of... 为……骄傲
Christmas card 圣诞卡片	injection 注射	officially listed 正式上市	announce 宣布
KPI 关键业绩指标	miracle 奇迹	NASDAQ 纳斯达克	make a toast 举杯发言

Ⅱ Read the Material Below and Answer the Questions

George: Good evening, Dr. Monroe (Moller Monroe). Thank you for inviting us to this ceremony.

Moller: Ah, George! My favorite student. How could I forget you! You were always asking me all kinds of challenging questions when you were a student in the college.

George: And this is my wife, Jennifer. I don't know if you remember her.

Moller: Of course, the Jennifer who is always sending me Christmas cards and Easter chocolates! I should call you Mrs. James now!

Jennifer: Congratulations on your newest patent in the insulin injection! It's such a miracle that you have done in the pharmaceutical field! My colleagues say that your patent is going to be nominated for the next year's Noble Price!

Moller: Well, thank you, Jennifer! By the way, I have heard some good news from George, right? Why don't you tell us?

George: Oh! It was nothing compared to yours, sir. This is the 5th patent you have invented.

Jennifer: Honey, you should definitely tell Dr. Monroe! Our company is officially listed on NASDAQ this Monday!

Moller: Seriously? Good on you George! I've always been proud of you! Come now; let me announce this great news to our guests! Ladies and gentlemen, may I have your attention. Here, I would like to give a special toast to my student Mr. George James.

Questions

1. What's the relationship between the speakers? (Monroe, George, Jennifer) Answer the question using adjective clause.

2. What have Monroe and George done that worth celebrating (值得庆祝)?

(注意:文中的George全名为George James,所以他的夫人随他姓James,为 Mrs. James)

Ⅲ Core Knowledge of Giving Praise

1. 祝贺某人。

Ways of Congratulations			
Good on you! Good work!	So you say you have won our company's soccer game? Good on your!	Congratulations!	Congratulations Mr. Sanders on your newest invention.
Well done!	Well-done Jonathan on your KPI this week.	It is a miracle! (which sounds impossible)	Dr. Harris, this project is a miracle!
Great work!	Great work, guys! You should definitely get some rest!	Awesome job!	Awesome job, Francis! Now the pipe stops leaking!
Proud of you!	Dimitri, you have broken the sales record! We are so proud of you!	Excellent work! Fantastic!	Your company has delivered some excellent work these days!

2. 长期持续的习惯。

be always doing	He is always calling his clients on their birthdays, and he is always seeking for new opportunities every minute in his life.
always been	My procurement manager has always been careless on her works!

Ⅳ Exercise and Activity

Imagine that your partner was a famous person, interview him/her, ask about his/her good habits, and congratulate him/her on his/her recent work, and make a role play out of it.

an engineering expert who is awarded the best employee of the year	
an athlete who just won the competition	
a famous writer who wrote a popular novel	
a chef (大厨) who just cooked something delicious	

(趁热打铁,课后练习见148页)

八 Appendix Easy Tips for Eating in a Restaurant 餐厅用餐攻略

1. Different food.

Appetizer	Drinks	Main Course	Deserts
vegetable salad 蔬菜沙拉	wine 葡萄酒	steak 牛排	ice cream 雪糕
cold ham with cheese 冷火腿加芝士	soda 苏打水	pizza 比萨	apple pie with cream 苹果派加奶油
fruit salad 水果沙拉	orange/apple/carrot juice 橙子/苹果/萝卜果汁	spaghetti 意面(细面)	cherry cake 樱桃蛋糕
bread crumb 面包屑	champaign 香槟	cream mushroom soup 奶油蘑菇汤	chocolate muffin 巧克力马芬
pudding 布丁	ice water 冰水	fried fish and chips 炸鱼薯条	pumpkin pancake 南瓜饼

2. Eating steaks.

Blue	0
Rare	15%–30%
Medium Rare	30%–40%
Medium	50%–60%
Medium Well	60%–70%
Well Down	70%–100%

3. Useful conversation methods in a restaurant.

When you invite someone to dinner 邀请某人吃晚餐	When you greet them in the restaurant 在餐馆和对方打招呼
Listen, I know this really good restaurant. Do you want to come?	Hey! How are you doing Jackson, congrats on your new job!
Have you tried Chinese hotpot (火锅) before? Let's go there this Saturday.	Good evening Mr.&Mrs. Richardson!
So why don't we grab a drink in the club tonight?	So, seeing anything nice to eat?
I would like to invite you to our house for dinner.	Any recommendations (推荐菜) for today's dinner?
Any plans after work?	

4. Invite a friend to the restaurant.

<table>
<tr><td rowspan="2">Do you have any plans after work?</td><td>Yes.</td><td>You want to grab a drink after work?
You want to grab a bite（吃点东西）after work?</td></tr>
<tr><td>No.</td><td>Well, if you got time next time, we can have a drink/dinner together; I know a really good place.</td></tr>
<tr><td colspan="3">I know a really nice restaurant down the block; we can give it a try.</td></tr>
<tr><td>When and where should we meet?</td><td>Is ________ o'clock OK for you?</td><td>You want to bring your wife with you? I'm bringing mine.</td></tr>
<tr><td colspan="3">I will be there, meet you there. So, are you allergic（过敏）to anything, or are there anything you don't eat?</td></tr>
<tr><td colspan="3">This is my number; call me if you are there.</td></tr>
</table>

Answers

Chapter A　Eating in a Project

Ⅱ Read the Material Below and Answer the Questions

1. They finally could eat something other than takeaways.

2. The water from the river is vastly contaminated, and the production line of the old factories has caused a lot of air pollution for the past 22 years.

Ⅳ Exercise and Activity

1. What food are you going to put in your home fridge? Fill in the fridge below.

a can of soda	two packs of chips
two boxes of coffee	a carton of eggs
three bars of chocolate	five bags of peanuts
a big chunk of beef	a few slices of cheese
four loaves of bread	some tomatoes

Chapter B　Socializing

Ⅱ Read the Material Below and Answer the Questions

1. There is supposed to be an appointment between her company and P&G tomorrow.

2. Betty was supposed to buy a new dress for the company's annual reception party.

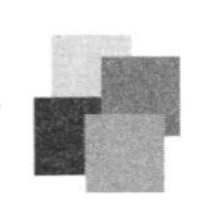

Ⅳ Exercise and Activity

1. 用am/is/are supposed to 和 was/were supposed to 来完成以下句子。

There is a school interview supposed to be here this Sunday.
Our materials were supposed to be delivered now.
Our facility management team is supposed to come next Monday.
The technicians from SHELL was supposed to send us copies of instructions and was going to organize an open discussion towards the process issues of our project, but they called in the last minute and told us that they couldn't come!

2. 用am/is/are supposed to 和 was/were supposed to 来回答以下问题。

I was supposed to get to work at 8:00 a.m., but I didn't.
There is supposed to be a new construction convention coming up next week.
Our annual meeting is supposed be in Shanghai.
I'm supposed to see my father today.
I was going to get some photocopies, but I was too busy to do so.
The Christmas party is supposed to be held on the 8th floor today.

Chapter C Congratulations

Ⅱ Read the Material Below and Answer the Questions

1. Dr. Monroe is the professor who taught George and Jennifer in the University.

2. Dr. Monroe's patent is going to be nominated for the Noble Prize, while George's company is officially listed in the NASDAQ.

Ⅳ Exercise and Activity

an engineering expert who is awarded the best employee of the year	Wow, good on you! I guess I'm going to work harder next year! Congratulations! You are truly an aspiration to us!
an athlete who just won the competition	Cool! You are so fast out there!
a famous writer who wrote a popular novel	That's a really fantastic book you have written!
a chef who just cooked something delicious	Man, that is delicious!

Class 9

Procurement
采购

Chapter A Procurement 采购

I Key Words and Expressions

installation materials 安装材料	defective materials 报废材料	paint 油漆	final offer 最终价	transportation fee 运输费
fire-proof materials 防火材料	artificial materials 人造材料	steel pipe factory 钢管厂	insufficient funds 资金不足	panel wall 面板墙
plastic materials 塑胶材料	insulating materials 绝缘材料	metal product plant 五金工厂	reduce some price 减价	How much are you buying? 你要买多少?
substitute materials 替换材料	road materials 筑路材料	concrete part factory 混凝土构件厂	minimum price 最低价	insane 疯狂的
rust-resisting materials 防锈材料	epoxy materials 环氧材料	current market price 当前市场价	structural materials 结构材料	order 订单

Ⅱ Read the Material Below and Answer the Questions

Dennis: Morning, Mr. Charles. Welcome to the F&P Construction Material Company. What can I do for you today?

Charles: Well yes Mr. Dennis, I need to see Mr. Finch to discuss some new purchases.

Dennis: Did you make any reservations?

Charles: As a matter of fact I did last week.

Dennis: OK, let me check and call him over. Please wait a moment.

Finch (F&P's manager): Hi Mr. Charles, good to see you! What do you need?

Charles: Well, I need to purchase a bunch of installation materials for our new project. I was wondering if you could show me around. Here is the list of items I need.

Finch: Well, my pleasure. Please follow me.

Charles: So tell me about these pipes? Are they made of fire-proof materials? How much are they?

Finch: They are fire-proof indeed. The current market price is 106 dollars per meter.

Charles: What about these rust-resisting pipes? How much are they?

Finch: They are 195 dallars per meter.

The transportation fee is on us. Also, they are made of PVC materials.

Charles: Em...you see, I think the price here is a little too high. I was thinking about 100 for these pipes. You see, I just went to the K&K steel pipe factory, and the price is 120 dollars higher than in 2016. Can't you give us some discounts?

Finch: That's pretty tough Charles. You know we have been cooperating for many years, and we only provide the quality guaranteed materials. The price is due to the current market. Our profit is very low.

Charles: I know Finch, but our company right now is quite insufficient on the funds. What's the minimum discount rate can you give us?

Finch: It depends on how much you want to buy.

Charles: We are purchasing at least 3,000 meters of pipes. Half fire-proof, halfrust resisting pipes.

Finch: All right. How about reducing the total price by 5%?

Charles: Could you make it into 15%?

Finch: 15% reduction on price! You must be kidding me! This is insane!

Charles: Look, Finch,this is quite a large order, and we are also going to add some other installation materials later. It would be best if you thought it over.

Finch: 10%, it's my final offer. The demand is pretty heavy nowadays.

Charles: Oh well, I guess we have a deal. You'd better deliver them to the site on time.

Finch: You could count on us! We will supply them without any delay.

Questions

1. What is this conversation about?
2. Do they reach an agreement? What's Finch's final offer?

Ⅲ Core Strategy of Telling People About Your Considerations and Requests

was/were considering	We were considering of a 20% reduction on the price of this material.
was/were thinking	Mr. Boston, I was thinking if I could take a vacation for 5 days.
was/were wondering	I was wondering if you could cover me up for a few minutes.
was/were looking for	It seems that 680 dollars a meter is too expensive for us, and we are looking for something around 500 dollars a meter.
was/were wondering, was /were thinking能让对方感觉到你的请求和决定是经过深思熟虑的，能让请求不那么直接。	

Tip:可以用以下的短语来吸引对方的注意力,然后将重要的观点告知对方(通常look, you see, listen会出现在key opinion之前)。

Look...	Look, the price of this artificial material is way too low.
You see...	You see, this product has large demands nowadays.
Listen...	Listen, our budget is limited (有限的), so we couldn't afford such an offer.

Ⅳ Exercise and Activity

1. 请填充下方空格,让对话更顺畅。

A: Randy,What do you need?

B: ____________ I could take a vacation for a week.

A: ____________ I could get a raise on my salary.

B: Well, we still need to consider about it first.

2. 请用I was wondering if/thinking if/considering if/looking at...来请求对方同意。

In the Office	
ask for help with some documents	
ask for a vacation	
ask for a 1,000 dollars raise	
In the Project Site	
ask for 20 more EHS staffs	
ask for more time for the drawings to be done	
replace the project manager	

Chapter B Procurement Preparation 采购准备

Ⅰ Key Words and Expressions

modification 优化	function 功能	machine 机械	procurement list 采购清单
upgrade 升级	institute 机构	earlier 更早	quality assurance 质量保障
sustainable 可持续的	precision 精确	pressure level 压力等级	convenient 方便的
admit 认可	accuracy 精准	humidity level 湿度	at once 立刻
win-win 双赢	EER (energy efficiency ratio) 能效比	package 包,承包	Let's see how it goes. 试试看。

 Ⅱ Read the Material Below and Answer the Questions

Davidson: Welcome to the SF-4 project site, Mr. Dawson. Also, it's a pleasure for me to show you around.

Dawson: Go ahead, Mr. Davidson. I have seen the modifications on the drawing. It seems that you have made some upgrades on the installation. Could you show me around?

Davidson: We have, actually and certainly. Please have a look. These are our newly modified pipes. Type A pipeline is longer and more sustainable than the type B pipes.

Dawson: Well, Mr. Davidson, to be honest, I still think that the type A's price is not as competitive as type B's price.

Davidson: Em, if you were only looking at the price, we admit that type A is more expensive than type B, but it seems to us that the quality assurance of type A is much better.

Dawson: I hope that it won't affect the function of the building.

Davidson: It absolutely won't! As a matter of fact, it upgrades the function of the building. Also, type A is a local brand, so I think it would be more convenient to pass the authorization, and the delivery of type A pipes would be faster than type B, while the transportation fee would be lower.

Dawson: Seems like a win-win deal. Let's see how it goes.

Davidson: I will leave the data on the desk.

Questions

1. What are the speakers talking about?
2. Which type of pipe did Mr. Davidson recommend? Why?

 Ⅲ Core Knowledge of Comparison

Adjective	Comparison	Adjective	Comparison	Adjective	Comparison
early	earlier	bad	worse	interesting	more interesting
low	lower	good	better	competitive	more competitive
high	higher	much	more	independent	more independent
strong	stronger	far	further, farther	reliable	more reliable
单音节词末尾加 er 如果该词以 y 结尾，去掉 y，加 ier		非规则形容词比较级		如果该形容词是多音节词，可以用 more 和 less 将其转变为比较级	

as + *adj./adv.* + as
The product quality of UTP equipment is just as reliable as the KNP's equipment, but not as expensive.
not as + *adj./adv.* + as
The director of KNP is not as experienced as the UTP's director.

Tip：在商务沟通，特别是比较不同产品的性能、价格的时候，我们经常会使用到比较级。

 Ⅳ Exercise and Activity

1. 阅读对话，将形容词变为比较级，然后与搭档进行对话。

Molly: So David, this is the procurement list that our suppliers delivered last Tuesday.

David: Alright, let's see. The machine from the KNP institute. So what's the precision of it, and the price? Tell me about it.

Molly: The precision and the accuracy of this machine is __________ (good) than UTP's, and it's 1,000 dollars __________ (cheap) than the UTP machine. By the way, our supplier Mr. Kim told me that they could deliver them 5 days __________ (early).

David: But the warranty of KNP machine is 12 months ________ (short) than the UTP machine.

Molly: Well, if you ask me, you also have to consider about the EER (energy efficiency ratio) level. UTP is not doing __________ (as...as) the KNP's machine.

David: Really? I thought they are the same. Excuse me, I've got an important call.

(phone ringing)

David: Hello, David Baker CCOC speaking.

Jerome: David, KNP has given Molly 200,000 dollars to recommend them in the project. The price should be _________ (low), and the pressure level is _________ (not as... as) the UTP machine which is also unauthorized by the government.

David: Oh! That's why! Molly, in my office please!

2. 用比较级比较下列商品和品牌。

A.

B.

C.

Reading a Graph 图表阅读

 I Key Words and Expressions

main force 主要力量	sanitary floor 洁净地板	curve graph 曲线图	ultrapure water 超洁净水
highly trained 高度熟练的	panel wall 洁净墙	ratio 比例	program 项目
labor market 人力市场	generatrix 母线	straight from 直接从	self development 个人发展
laid off 失业	pie graph 饼图	seattle 西雅图	survey 调研
financial disputes 经济纠纷	histogram 柱状图	immature不成熟的	resigning 辞职

II Read the Material Below and Answer the Questions

Dante: So Mr. Daniell, it looks that you have completed the recruitment for the project. Why don't you share some details with me?

Daniell: Well, you can take a look at the graph, Mr. Dante.

Daniell: As you can see on the graph, the main labor force of our ultrapure water program this year are mostly young recruits with only 1–2 years of experience. The second largest forces are workers with 3–4 years of experience, and it covers 34 percent of our total recruitment.

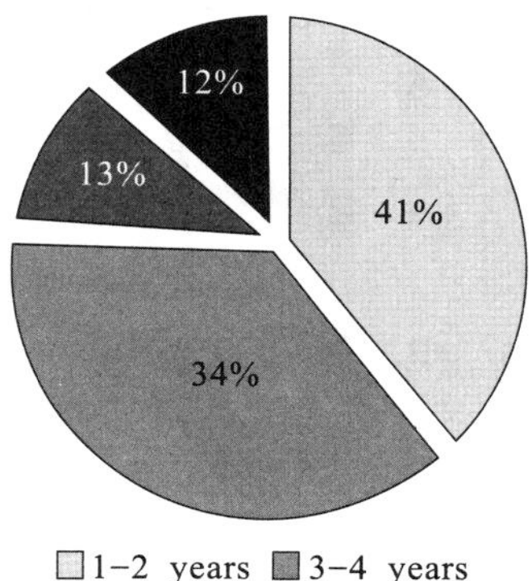

Dante: It's quite a good job you have done there Mr. Daniell, but you do understand that this project needs many workers with at least five years of experience.

Daniell: To be honest, we have met some difficulties this year while recruiting highly trained/experienced workers due to this year's high labor market price. However, the good news is that our rival company STPI laid off many of their skilled workers last month due to some financial disputes. So although only 25 percent of our recruits this year have 5 years or above experience, 12 percent of them have over seven years of experience, while most of them were recruited straight from STPI.

Dante: Thank you, Mr. Daniell. It looks like you have done a pretty good job this year.

Questions

1. What is this graph about?
2. Is Mr. Dante satisfied with Mr. Daniell's work? Why?

Ⅲ Talking About a Data or a Graph

1. 曲线图(curve graph).

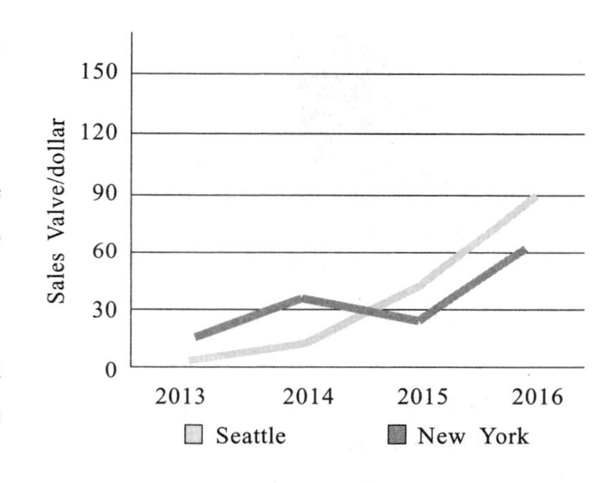

第一步，我们在讲解曲线图的时候通常会先介绍曲线的用途。比如说右侧的曲线图是关于两个城市的销售额。

So this graph shows the sales value of Seattle and New York from 2013 to 2016.

第二步，讲讲图的起点2013年的情况。

We started small in New York and Seattle; at the beginning our sale value was about...

第三步，讲解趋势特点，上升、下降、波动还是震荡？为什么？

The graph shows that the sale value was gradually increasing with slight turbulence in 2014-2015 in New York, because our rival company made a huge reduction on the price (reason).

第四步，结论。综合整个曲线图你能发现什么特点呢？又有什么计划呢？

It seems that the sales value in Seattle and New York has been growing healthily, with only a little setbacks. We are going to reach 150 million by the end of next year.

2. 饼图(pie graph).

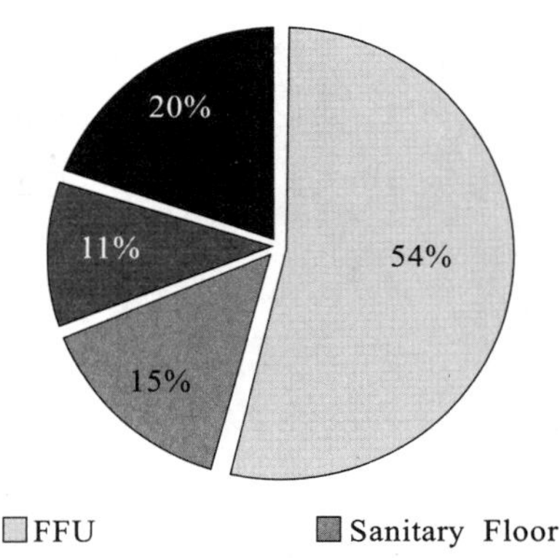

第一步，先介绍饼图的用途。比如，右图是为了展示年度的采购支出。

The graph on the right shows the ratio of our annual procurement.

第二步，陈述图中比例的特点和其中的原因。

The major procurement this year is the FFU system, which covers 54% of our total budget, because... 20% is from the generatrix...

第三步,结论。综合整个饼图你能发现什么呢?又有什么计划呢?

Tip:不要仅仅描述图中的数据本身,要多分析数据背后的原因。

3. Useful expressions.

increase 增长	percentage 百分比	turbulence 波动	have/has potential 有潜力	steady 稳定的
decreasing 降低	rapid development 飞速发展	mostly 大部分	started small 从小开始	continuing 持续
reduction 减少	major loss 重大损失	rarely 极少	growing strong 增长势头强	recess 停滞

Ⅳ Exercise and Activity

请尝试描述下图的内容。

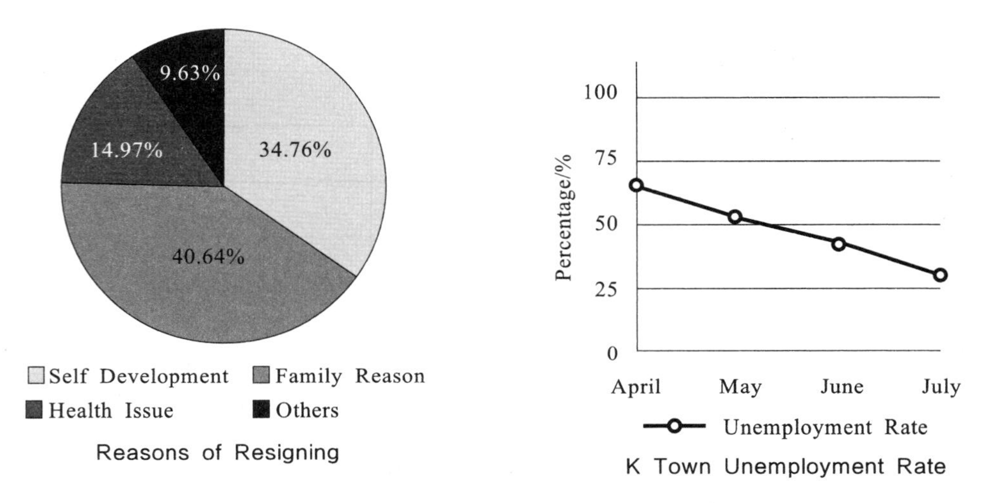

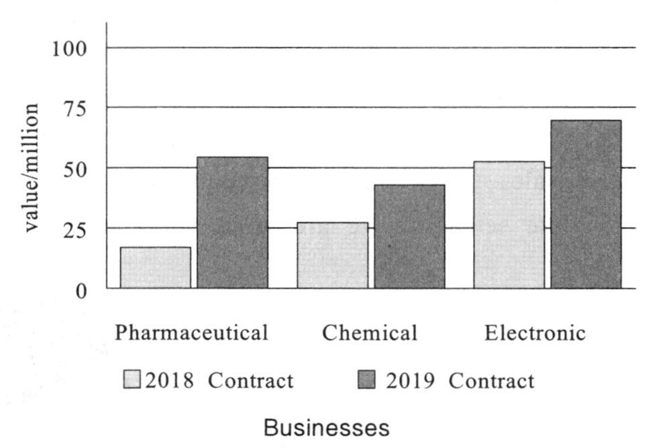

(趁热打铁,课后练习见149页)

Appendix Engineering Machines and Vehicles 建筑器械与车辆

road machinery 道路机械	stabilized soil mixer, road cutter, road marking vehicle, snow remover, asphalt distributor, asphalt paver, pavement maintenance vehicle, pavement breaker, road block removal truck, garbage truck
digging machinery 挖掘机械	excavating loader, hydraulic wheel excavator, trencher
concrete machinery 混凝土机械	concrete mixer, concrete pump, concrete breaker, concrete paver, concrete pump truck
hoisting machinery 起重机械	windlass, truck crane, crawler crane, elevator, lifter, pipe crane, aerial platform vehicle, tower crane, towing machine, forklift, gantry crane
earth moving machinery 土方机械	transport vehicle, scraper, grader, dumper, tyre loader, crawler loader, bulldozer, dump truck, conveyor
compaction machinery 压实机械	vibrating tamper, vibratory roller, static roller, compactor
drilling machinery 打钻机械	heading machine, multifunctional drilling rig, rotary drilling rig, percussive drilling rig, construction drill
decoration machinery 装修机械	electric nacelle, mud pump, mortar pump, winch, heater, edger, sprayer pump, plastering machine
pile driving machinery 桩工机械	pile hoop, pile frame, pile driver, hydraulic shear, diaphragm-wall grabber
pre-stressed machinery 预应力机械	hydraulic tong, reinforced steel adjusting cutter, jack, anchor device, hydraulic machine
rock drilling machinery 凿岩机械	compressor, rock drill, breaking hammer
Must learn vocabulary(必学词汇): mixer, remover, paver, breaker, crawler, lifter, crane, loader, pump, drill.	

Do you know what they are?

Answers

Chapter A Procurement

Ⅱ Read the Material Below and Answer the Questions

1. The speakers were talking about which type of pipes they should choose, and the price of it. Charles is looking for some discounts.

2. They did reach a final deal, 10% discount.

Ⅳ Exercise and Activity

1. 请填充下方空格,让对话更顺畅。

A: Randy, What do you need?

B: I was thinking if I could take a vacation for a week.

A: I was thinking if I could get a raise on my salary.

B: Well, we still need to consider about it first.

2. 请用I was wondering if/thinking if/considering if/looking at...来请求对方同意。

In the Office	
ask for help with some documents	I was wondering if you could help me with some documents. You see, the documents have been piled up（堆积）for days!
ask for a vacation	I was wondering if you could let me go for a vacation.
ask for a 1,000 dollar raise	I was thinking of getting a 1,000 dollars raise.
In the Project Site	
ask for 20 more EHS staffs	I was wondering if you could send 20 more EHS staffs on site.
ask for more time for the drawings to be done	I was thinking if you could give us 10 more days for the drawings to be done.
replace the project manager	I was considering changing our project manager, because he is just too unreliable.

Chapter B Procurement Preparation

Ⅱ Read the Material Below and Answer the Questions

1. They are discussing about the materials and the procurement.

2. David recommended type A material, because it would actually upgrade the function of the building. Also, type A is a local brand, so it would be more convenient to pass the authorization, and the delivery of type A pipes would be faster than type B, while the transportation fee would be lower.

Ⅳ Exercise and Activity

1. 阅读对话,将形容词变为比较级,然后与搭档进行对话。

Molly: So David, this is the procurement list that our suppliers delivered last Tuesday.

David: Alright, let's see. The machine from the KNP institute. So what's the precision of it, and the price? Tell me about it.

Molly: The precision and the accuracy of this machine is better (good) than UTP's, and it's 1,000 dollars cheaper (cheap) than the UTP machine. By the way, our supplier Mr. Kim told me that they could deliver them 5 days earlier (early).

David: But the warranty of KNP machine is 12 months shorter(short) than the UTP machine.

Molly: Well, if you ask me, you also have to consider about the EER (energy efficiency ratio) level. UTP is not doing as well as (as...as) the KNP's machine.

David: Really? I thought they are the same. Excuse me, I've got an important call.

(phone ringing)

David: Hello, David Baker CCOC speaking.

Jerome: David, KNP has given Molly 200,000 dollars to recommend them in the project. The price should be lower (low), and the pressure level is not as high as(not as....as) the UTP machine which is also unauthorized by the government.

David: Oh! That's why! Molly, in my office please!

2. 用比较级比较下列商品和品牌。

A.

Windows 10 is more convenient and high-tech than Windows 95. You see, its CPU is so much stronger than Windows 95. Window 95 is just not as user-friendly as Windows 10.

按照此例回答其他问题即可。

Chapter C Reading a Graph

Ⅱ Read the Material Below and Answer the Questions

1. This graph shows the recruitment data of the project.

2. Mr. Dante is pretty satisfied with Mr. Daniell's work because Mr. Daniell has hired many experienced workers for the project.

Ⅳ Exercise and Activity

请尝试描述下图的内容。

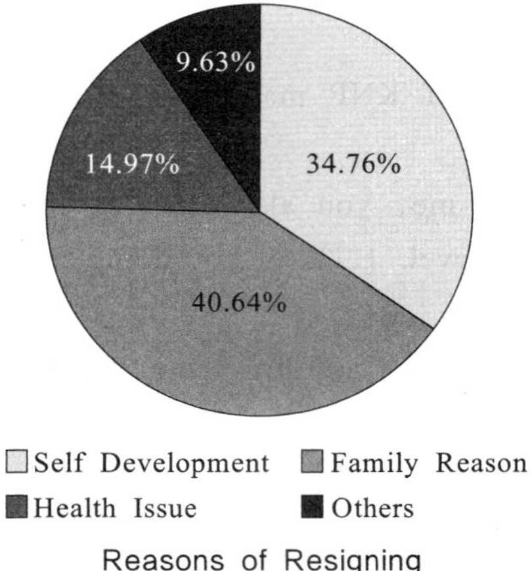

Reasons of Resigning

This graph shows the reasons of resigning. After our survey, we found out that 41 percent of them left us for family reasons. 35 percent of them left for further career development, 15 percent for their health issues, and only 9 percent for other reasons.

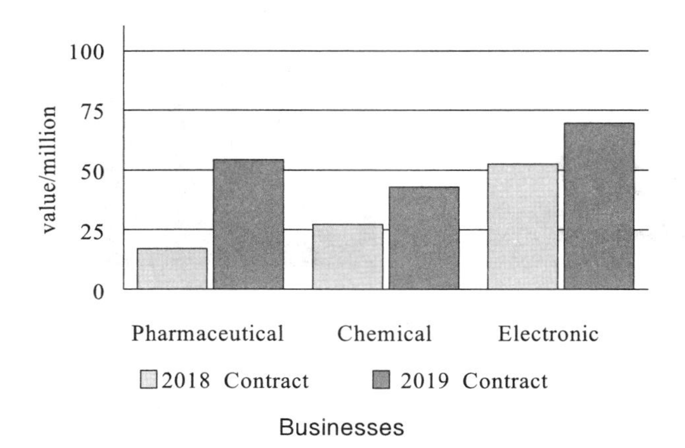

Businesses

The graph above tells us the contract growth of our company's pharmaceutical, chemical, and electronic businesses.

As for the pharmaceutical business, we have been faced with some difficulties and have dropped 15 million dollars on sales. However our chemical and electronic contracts have obviously (明显地) been increasing.

Promises and Regulations
承诺与规章制度

Chapter A Promotion and Promises 升职与承诺

 I Key Words and Expressions

sales engineer 销售工程师	open up the markets 打开市场	you know what I mean 你懂的
limited time 有限时间	team building 团队建设	company's culture 公司文化
make a deal 达成交易	family day 家庭日	worker's loyalty 员工忠诚度
pull it through 跨过难关	stuff like that 类似的事情	hopefully 但愿
manage 处理	great promises 前景可人	appoint 任命

 Ⅱ Read the Material Below and Answer the Questions

Adam: Morning, Mr. Olsen. Sorry for keeping you waiting, please have a seat.

Olsen: Morning, and thank you for the promotion opportunity, Mr. Adam.

Adam: It says on the files that you have been working as a sales engineer in our company for the past three years right?

Olsen: Well I have, and I have just made a deal with WEST.

Adam: So tell me, what do you think about this project?

Olsen: Actually, to be honest, I think the WEST project was challenging. We had to complete the project within a minimal time with extremely strict standards and things like that, but we managed to pull it through.

Adam: Have you made any changes to improve your workmates' efficiency and loyalty these three years?

Olsen: Absolutely, I have organized engineering training, English training, management training, and training as such to improve their skills and efficiency, also we are always holding many team buildings, family days, and events like that to spread our company's culture, as well as to influence their loyalty to CCOC.

Adam: What are you going to do after your promotion?

Olsen: The first thing I'm going to do is to open up the markets in India and sign our newest project in Iran. Hopefully, we could complete it on time.

Adam: Which field would you put your main focus on after you are appointed as the chief sales manager?

Olsen: If I were appointed, I would focus on the pharmaceutical markets. Because I can see great promises in this field.

Adam: Thank you, Mr. Olsen, this is all I need to know, and we will send you the result afterwards.

Olsen: Thanks again for the opportunity.

Questions

1. What exactly has Mr. Olsen done back when he was a sales engineer?
2. What are the things he is going to do after his promotion?

Ⅲ Facing an Interview

1. Simple verbs and continuous verbs.

Simple Verbs（一般动词）		Continuous Verbs（延续性动词）	
What do you think about this project?	It seems to me that this project still needs more supervision.	Are you working on the procurement lately?	As a matter of fact, I am working on it right now.
Have you completed any foreign contracts?	I have done 3 foreign projects.	How long have you been working in this industry?	I have been working in this industry for 7 years.
Did you meet Dr. Lee last night?	I did, and we chatted about our business.	What were you doing with Jenny last week?	We were staying at home watching TV.
What are you going to do for the next 5 years?	I'm going to make a million dollars.		

2. Vague expressions（模糊表达法）.

<table>
<tr><th>Normally</th><th colspan="2">Vague Expression</th></tr>
<tr><td rowspan="2">I have worked for Roche, Intel, BMW, AT&T, P&G, Lenovo, and SIEMENS...</td><td>+and…like that</td><td rowspan="2">I have worked for Roche and companies like that.</td></tr>
<tr><td>+and everything</td></tr>
<tr><td rowspan="2">I like to go to a club, drink some beer, join a party, stay with my friends... after work.</td><td>+and stuff like that</td><td rowspan="2">I like going to clubs and things like that after work.</td></tr>
<tr><td>+and you know what I mean</td></tr>
<tr><td colspan="3">and (companies, places, foods, sports, people, jobs, interests...) like that 意味着某个公司、某种食品、某种工作、某种爱好等。</td></tr>
</table>

模糊表达法能省略许多排比句中的元素，让句子不会太无聊。

Ⅳ Exercise and Activity

Please use the vague expressions to make plans for a team-building event in your company.

Who is going to be invited?	
What are we going to discuss about?	
What are we going to do in the evening?	

Chapter B Regulations 规章制度

I Key Words and Expressions

announce 宣布	instruction 指示	dismiss 解散	illegal 非法的
common sense 常识	dangerous 危险的	workstation 工位	license 执照
forbidden 禁止	reply 回复	openly 公开地	unrealistic 不切实际的
according to 根据	work related 工作相关的	intolerable 不可容忍的	keep... to themselves 保密
punish 惩罚	leak out 泄漏	acceptable 能接受的	chew 咀嚼

II Read the Material Below and Answer the Questions

Carl: Morning Gentlemen, I'm Carl Sanders, your new project director. It's a pleasure for me to work with you guys, and I wish that I could make a better tomorrow for each one of you here.

Mathew: Thank you, Mr. Sanders. As for today's meeting, Mr. Sanders is going to announce some new rules and regulations in our department as well as in a project.

Carl: Let's start with common sense. First, smoking on the site is forbidden. Second, not wearing the PPE according to the code will be punished. Third, one must strictly follow the instructions while operating dangerous equipment.

Mathew: All right, now for the second part, office rules.

Carl: All right, arguing openly in the office is forbidden. Immediately replying our client's email is something that we should always remember, and of course, watching non-work related websites is also intolerable in our office.

Mathew: If you don't mind Mr. Sanders, I would like to add a few more things.

Carl: Go ahead, Matt.

Mathew: Always remember to keep our business information to ourselves, and keep our patents (专利) from leaking out.

Carl: All right, that would be all, now meeting dismissed. Go back to your workstations.

Mathew: Thank you, sir.

Questions

1. What did Mr. Sanders talk about in the meeting?
2. What are the rules and regulations in the office?

 Ⅲ Core Knowledge of Talking About Rules and Regulations

动名词 v.-ing 作为名词	介词后的 not+doing/not+doing(动名词)
Eating on the subway is impolite.	It's illegal not bringing a license while driving.
Disobeying the law is unacceptable.	Not wearing a PPE while working is foolish.
Wearing PPE is the rule of our site.	
介词之后的 v.-ing	介词之后的not+to do
It's impolite to eat on the subway.	It's OK not to bring cash to the supermarket nowadays.
It's our rule to wear PPE on the site.	It's unrealistic to become successful quickly.

类似这样的动词做主语的情况一般是为了强调行为是否得当(特别是在制定规则的时候会用到)。

 Ⅳ Exercise and Activity

1. 将日常的口头提示改成行为规范。

When You Speak	In the Regulation
Jason, don't chew gum in the class.	Chewing gum in a class is rude.
Don't smoke in here Peter.	
Please check our equipment before you go.	
Adam needs to organize a meeting each morning.	

2. 平时在项目上和办公室中会有什么规则呢？请将它们列在下方表格中。

Office	Project

 Chapter C He Told Me That 他告诉我

 Ⅰ Key Words and Expressions

production center 生产中心	inform 通知,告知	on a strike 罢工
business card 名片	collect 收集	I would love to... 很希望……
send one's regards 问候你	rock bottom price 最低价	Whether you could... 不知能否……
on behalf of... 代表……	arrange 安排	percentage 百分比
workshop 车间	switzerland 瑞士	daily schedule 日程安排

 Ⅱ Read the Material Below and Answer the Questions

Gerald: Welcome to CCNPC, Mr. Smith. I'm Kevin Gerald, the supervisor of the production center. This is my business card.

Smith: Thank you, Mr. Gerald. Mr. Richardson sends his regards on behalf of KNP.

Gerald: Well, here is the daily schedule for Mr. Richardson. After Thursday's morning meeting, we are going to visit the workshop and have another meeting with Mr. Charles, our project director at the production center.

Smith: All right, I will inform Mr. Richardson. So, any plans on Thursday evening?

Gerald: As a matter of fact, Mr. Charles was wondering if you were free for dinner in the evening. Ah! Please do ask Mr. Richardson to bring Mrs. Richardson to come over, because Mrs. Charles will also be there.

Smith: OK, and Mr. Richardson wants to know when you will deliver us the samples of our equipment.

Gerald: Would next Friday be OK? We need some time to collect the information on the quality for a couple of days.

Smith: And the price, do we have a deal on the price Mr. Richardson offered earlier.

Gerald: Well, we discussed it last week. It seems to us that $30 per unit is our rock bottom price. If CCNPC would purchase more than 5,000 units, we may arrange a small percentage of discount. Would it be OK?

Smith: All right, I will let Mr. Richardson see to it. Thank you, Mr. Gerald.

Smith: Morning, Mr. Richardson. I have just talked to Mr. Gerald from CCNPC, and he has given me a full schedule on Thursday this week.

Richardson: Go ahead. I'm listening.

Smith: Mr. Gerald said that we are going to have two meetings on Thursday this week, and asked whether you have time for dinner with their project director. Also, he also wondered if you could even bring your wife to the dinner because he is bringing his.

Richardson: Tell him that I would love to, but my wife and kids are in Switzerland for vacation. And did you ask him about the price we offered?

Smith: I did, and he told me that 30 dollars per piece is his final bottom price, and he offered to give us a discount if we could purchase 4,000 units or more.

Questions

1. What is Mr. Richardson's schedule on Thursday?(Report it)
2. Did Mr. Smith make any mistakes or miss（遗漏）anything while reporting?

Ⅲ Core Knowledge of Reporting What Someone Said

Normally We Say...	Reported Speech（引语）
When will Mr. Dawson come to Berlin?	He asked when would Mr. Dawson come to Berlin.
The meeting is going to be on tomorrow.	She told me that the meeting is going to be on tomorrow.
Do I have any insurance?	She wanted to know whether she has any insurance.
My boss told Todd that our workers are on a strike.	My boss told Todd and he told me that our workers are on strike.
He/she said that/told me that/says that... 转述某人的话	
He/she wanted to know whether/if... 转述某人的问题	

Ⅳ Exercise and Activity

Read the dialogue, report it using reported speech.

Hi, Mr. Sanders. It's me, Mark. I'm calling today to thank you for the price and discounts you offered last Friday. It's delightful to speak with you during the meeting, and the offer seems to be an excellent match for our project. We wish to cooperate furthermore with MAT in the near future. Also, we want to invite you for dinner tomorrow evening at the Hilton Hotel. Could you bring your family with you? We would be happy to see them come.

（趁热打铁，课后练习见149页）

Appendix Contracts and Agreements 合同与协议

1. 各种不同的合同。

Agreement/Assignment/Contract of（协议/转让/合同）			
agreement of reimbursement	agreement on future delivery	contract of guaranty	contract of the transfer of technology
agreement of option	agreement of intent	contract of hire	assignment of policy
agreement of reinsurance	agreement on government procurement	contract of indemnity	assignment of interest
claim agreement	contract of services	contract of sale	contract of lease

2. 常见的合同内容。

title	price	inspection	arbitration
generals	insurance	confidential	force majeure
name of the commodity	shipment and delivery	guarantee	applicable laws
quality	industrial and property right	claims	miscellaneous laws
quantity	payment	breach/rescission of contract	witness

3. 自拟合同。

Try and create a business contract with different purposes (such as an employment contract, a cooperation contract, or a contract for services), and share it with your partners.

Answers

Chapter A Promotion and Promises

Ⅱ Read the Material Below and Answer the Questions

1. Mr. Olsen has made a deal with WEST, and has organized all kinds of trainings, team buildings to enhance the efficiency of the project.

2. He is going to open up the markets in India, Iran, and focus on the pharmaceutical businesses.

Ⅳ Exercise and Activity

Please use the vague expressions to make plans for a team-building event in your company.

Who is going to be invited?	We are going to invite our engineers, managers, and people like that to come.
What are we going to discuss about?	We are going to discuss about the newest project details and everything.
What are we going to do in the evening?	We are going to enjoy the evening by having a big group dinner and things like that.

Chapter B Regulations

Ⅱ Read the Material Below and Answer the Questions

1. What did Mr. Sanders talked about in the meeting?

2. What are the rules and regulations in the office?

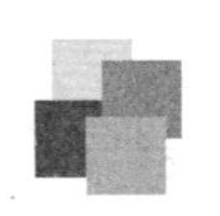

Ⅳ Exercise and Activity

1. 将日常的口头提示改为行为规范。

When You Speak	In the Regulation
Jason, don't chew gum in the class.	Chewing gum in a class is rude.
Don't smoke in here Peter.	Smoking in the office is restricted.
Please check our equipment before you go.	To check your equipment after work is common sense.
Adam needs to organize a meeting each morning.	Organizing meeting each morning is really important.

2. 平时在项目上和办公室中会有什么规则呢？请将它们列在下方表格中。

Office	Project
Arguing loudly in the office is inappropriate (不适当的).	Running around in the project site is really dangerous.
Talking behind people's back is not a good idea.	Following the instructions in the training is really important to every worker on site.
Eating in the office during work time is a violation of the office regulations.	Attending the morning meeting on time is the first rule on the site.

Chapter C He Told Me That

Ⅱ Read the Material Below and Answer the Questions

1. Gerald says that Mr. Richardson is going to visit the workshop after the morning meeting, and he is going to have an open discussion and a dinner with Mr. Charles.

2. Gerald says that they can get a discount if they bought at least 5,000 units (not 4,000).

Ⅳ Exercise and Activity

Read the dialogue, report it using reported speech.

Hi, Mr. Sanders. It's me, Mark. I'm calling today to thank you for the price and discounts you offered last Friday. It's delightful to speak with you during the meeting, and the offer seems to be an excellent match for our project. We wish to cooperate furthermore with MAT in the near future. Also, we want to invite you for dinner tomorrow evening in the Hilton Hotel. Could you bring your family with you? We would be happy to see them come.

Mark says that he wants to thank you for the price and discounts you offered last Friday. And he says that it's very enjoyable to speak with you, and wishes to cooperate further with MAT. Also, he asks whether you have time for dinner tomorrow, and if it's possible to bring your family to come over.

Meeting Minutes
会议纪要

Chapter A Weekly Meetings 周会报告

 Ⅰ Key Words and Expressions

regular meeting 例会	detailed schedule 详细日程安排	vendor 卖家
meeting minutes 会议纪要	remark 反馈,评论	qualified 合格的
crew 团队	materials for approval 待审核材料	task 任务
management company 管理公司	sample for approval 待审核样品	deadline 截止日期
contact list 联系人员名单	sort out 整理	conclusion 结论

 Ⅱ Read the Material Below and Answer the Questions

Steve: Good morning gents, thank you for attending the regular weekly meeting where we should discuss the issues from the meeting minutes.

Ray: Morning Steve, before we start, I would like first to introduce our crew from CCOC. Mr. Sanders our project manager, Mr. Lee, our procurement manager, and Mr. Martin, our EHS manager.

Bowen: And I'm Bowen, the representative for NDP Management Company, nice to meet you.

Steve: You too, now shall we begin. Let's take a look at item 1.01, the regular meeting schedule for the next two weeks. We have got no problem about that correct?

Item	Meeting No./Date	Importance(T/I)	General		Responsible
1.01	1/01.08.2019	I	regular weekly meeting on	Regular construction meeting at 9:00 a.m.	CCOC / (15.01.2019) (22.01.2019)
1.02	1/01.08.2019	T	list of contacts	CCOC's email should be sent to KDA.	CCOC/ (10.01.2019)
1.03	1/01.08.2019	I	emails	CCOC's email should be sent to KDA and the management group (NDP).	CCOC/ (10.01.2019)

(I=information, T=task)

Ray: Nope and 9:00 a.m. is quite reasonable.

Steve: All right, the next two items, the list of your contacts, did you prepare them and hand them to us?

Ray: We did, and we are going to add the email list as well this afternoon.

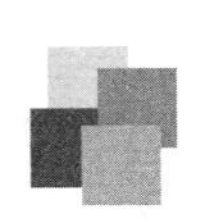

Steve: All right, let's have a look at key two, the schedules.

Steve: Ray, have you guys prepared the detailed schedules for the next three months?

Ray: Not yet, because the project has just been started, and we still need to look into many things.

Steve: When are you going to finish it?

Ray: Before the first of March 2019. However, we have handed in the first two weeks plan for the project.

Steve: Noted, and thank you. Now, Bowen, did NDP check the schedules?

Bowen: We did it yesterday, and made some remarks.

Steve: Please show them to us as well.

Bowen: Absolutely.

Steve: All right, and did NDP prepare a list of materials for approval?

Bowen: We are still waiting for CCOC's list of sample materials.

Steve: Ray, When will it be done, and why is it not done?

Ray: It will be done before next week's regular meeting, because we still need to sort out all the materials for you carefully, and we want them to be qualified for the project.

Steve: Thank you Ray, but the schedule is the schedule. What about the vendor's list?

Bowen: We have received it from CCOC last Friday, and are now checking on it.

Items	Meeting No./Date	Importance (T/I)	GC Contract Issues		Responsible
2.01	1/01.08.2019	T	detailed schedule hand in	(1) CCOC hand in the detailed schedule. (2) NDP should check the schedule.	(1) COC (2) NDP
2.02	1/01.08.2019	T	two weeks schedule check	CCOC should hand over the two weeks plan.	CCOC
2.03	1/01.08.2019	T	main materials and equipment	(1) Prepare the materials for approval (2) List of all vendors.	NDP

(T=task)

Questions

1. When will be the next regular meeting?
2. Why isn't the list of samples ready yet?

Ⅲ Core Knowledge for the Meeting Minutes

每周的例会是项目上不可或缺的工作，所以作为一名合格的工程事业工作者，能够读懂会议纪要并且给业主相应的回答是必须掌握的技能。

Minutes of the Construction Meeting for KDA(业主名称)

I=information, T=task, D=deadline, C=conclusion

信息，任务，截止日，结论（有的必须完成，有的仅需提供相关信息更新即可）

Meeting	Content		Status
item/meeting No./date	type/keywords	description/task/conclusion/statement	responsible/deadline/finalized

其中I(information)仅需提供相应信息，
而T(task)作为任务需要按期完成

具体怎么做和分工。并不是所有的问题都需要回答。如果作出改动，及时订正。

Item	Meeting No./Date	Importance(T/I)	GC Contract Issues		Responsible
2.01	1/01.08.2019	T	detailed schedule hand in	(1) CCOC hands in the detailed schedule. (2) NDP should check the schedule.	(1) CCOC (2) NPPC (the management group)
2.02	1/01.08.2019	T	two weeks schedule check	CCOC should hand over the two weeks plan.	CCOC
2.03	1/01.08.2019	T	main materials and equipment	(1) Prepare the materials for approval (2) List of all vendors.	(1) CCOC 02.08 (2) NDD 02.08.2019

时间是相当关键的，约定的时间要具体并且可行，也可以以工作日作为时间单位。

Tip: There would be more than two parties in a meeting, identify them. 会议上不仅仅只有甲乙方，必须要提前知道有几方的代表参与。

Ⅳ Exercise and Activity

Check out the meeting minutes in the Appendix of class 11. Divide into groups

of three students, and practice handling a regular meeting by playing different sides of the parties. (Note down all the deadlines.)

Chapter B Delayed 项目延期

Ⅰ Key Words and Expressions

excavation 挖土	cave 洞穴	millions 几百万
excavation vehicle 挖土车辆	landslide 滑坡	severe rain 大暴雨
submit 提交	far more complicated 复杂多了	unacceptable 不可接受的
original schedule 初计划	excuses 借口	previous 之前的
geological survey 地质调研	treatment report 解决方案	pump concrete 注入混凝土

Ⅱ Read the Material Below and Answer the Questions

Steven: Thank you for coming, Mr. Ray. What we are going to do today is to discuss a few keys towards the issues of delay.

Ray: Please go ahead.

Steven: On Item 2.01, you told us two weeks ago that the excavation should begin on the 21st of January 2019, but still I can't find any excavation vehicles around the site. What's wrong with it?

Ray: We still need to submit the approval list and the contract to let them enter into the site.

Steven: But it would delay our original schedule. It's 5 days over the deadline!

Ray: This problem was also caused by the delay of approval and the geological survey. We are going to push them, and hopefully they can start their work on the 28th.

Steve: Now, about the geological survey, when will it be finished, and when will you hand in the treatment report?

Ray: Well, for your information, the survey has been done this morning, and the result is far more complicated than we expected, so the previous solution of pumping concrete into the caves might be changed. So we need 2 more days to analyze them.

Steve: Don't tell these excuses to me, and also, why hasn't the temporary office been completed yet?

Ray: That's not our fault, because our purchase has been made, but the suppliers didn't deliver them.

Steven: Why and how?

Ray: They said it's because of the severe rain, and they told me that it would take about another 3 days to deliver them, and we also need another three days to assemble them. So it will be finished before the second of February.

Steven: Wait, what. This is unacceptable! Each day of delay is going to cost us millions!

Ray: Unfortunately this is what we are facing right now.

Questions

1. What are the speakers talking about?
2. What are the delays? And why?

 Ⅲ Core Knowledge of Mentioning Delays in a Meeting

This is a list of delays, with the original schedule.

Items	Meeting No./Date	Importance(T/I)	Delays		Responsible
2.01	2	T	excavation (delayed)	The excavation should begin before...	CCOC 21.01.2019
2.02	2	T	ground survey (delayed)	The geological survey report should be delivered before...	CCOC 25.01.2019
3.01	2	T	the temporary	Building hasn't been finished yet.	(1) CCOC (2) NDP

Before negotiating the delays, you must note down the situations at hand.

Items	Meeting No./Date	Importance(T/I)	Plans to Cope with It		Responsible
2.01	2	T	excavation (delayed)	The excavation approval was not approved yet.	CCOC 28.01.2019
2.02	2	T	ground survey (delayed)	There is a new issue to the caves, and the previous plan might be changed.	CCOC 27.01.2019
3.01	2	T	the temporary office (delayed)	The suppliers delayed the delivery due to severe rain.	CCOC 02.02.2019

reason changed schedule

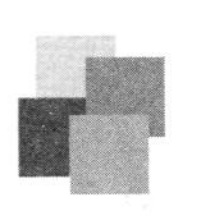

Explaining 解释

Ⅰ Key Words and Expressions

urgent situation 紧急情况	unknown 未知的	severe rain 大暴雨
push the process 推动进程	agreed on 就……达成共识	out of stock 售完(缺货)
site surroundings 现场周边	substitute vendor 备用卖家	exhausted 疲惫不堪的
reanalyze 重新分析	region 地区	staff replacement 人员调动
specification 规格,详述	violate 违反	operation mistake 操作失误

Ⅱ Read the Material Below and Answer the Questions

Stent: Hello Mrs. Aniston, sorry to bother you. We have got an urgent situation to report. Also, the representative for CCOC is here to provide us with more information.

Aniston: Well, I'm listening.

Stent: Mrs. Aniston is the CEO of KDA, and you can explain to her about the situations in the project.

Aniston: Thank you! Due to the report, I can see that the excavation is going to be delayed for another week.

Read: It is. It's because the approval has't been given to us yet by the NDP. So we cannot sign the contract on time.

Aniston: Mr. Read, I want no excuses. I need solutions!

Read: As I have discussed with my colleagues, they told me that they had made several phone calls to push the process, and it's going to be done this afternoon, and the excavation vehicles will arrive tonight.

Aniston: OK, that sounds all right. What about the ground survey? What's wrong with it?

Read: The geological survey team has discovered the total size of the underground cave. It covers 14% of the entire area. Also, the severe rain has caused a landslide near our site. So the cave treatment must be reanalyzed and changed.

Aniston: So it's going to be another two days for the report, and how long is it going to take for you to start the treatment?

Read: It's hard to say because as for matters underground, we are looking at something unknown, so more surveys need to be done.

Aniston: When?

Read: The treatment would start as soon as KDA approves the report. So, it will be on the 29th of January.

Aniston: So that's one thing, but what about the temporary offices? Still behind?

Read: The severe rain mainly causes it...

Aniston: Stop right there, enough of these excuses. It's said on our contract, rainy weather should not affect the schedule we agreed on.

Read: It is, but as for the current situation we have contacted our headquarter to provide materials from substitute vendors to deliver documents from other regions, and we are all working overtime with extra workers to make sure that it won't affect the entire schedule.

Aniston: Remember not to violate our contract.

Question

Would the entire schedule be delayed? Why?

 Ⅲ Situations You Might Encounter in a Project

Weather/Nature 天气/自然	Suppliers/Vendor 供应商	Labor 劳动力	Owner 业主
severe rain heavy storm winter chill summer heat minor earthquake humidity/flood wind	late on delivery price change quality issues out of stock late on approval	lack of labor not enough training operation mistakes exhausted workers accidents overtime labor	investment delay drawing changes equipment changes staff replacement arguements

关于工期常用的词组：

behind 延后	close 临近	inevitable 不可避免
near 接近	next 下一个	violate 违反
ahead 提前	besides 另外	time extension 时间延期
above 超出	tied up 无所适从	overtime work 加班
under 以内	exhausted workers 疲惫的员工	experts and specialists 专家团队

 Ⅳ Exercise and Activity

假设在项目上因遇到了以下问题造成延迟，你会怎么做？

The workers have been working overtime on site for a week! Still the schedule is behind!

The delivery is late, and the building hasn't been started yet!

The owner wants to change our project manager.

Waiting for government approval.

Quality issues with our goods.

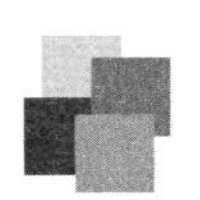

Appendix Meeting Minutes Sheets 会议记录表

分成三个小组讨论以下表格。

Items	Meeting No./Date	Importance(T/I)	GC Contract Issues		Responsible
2.01	1/01.08.2019	T	detailed schedule hand in	(1) CCOC hands in the detailed schedule. (2) NDP should check the schedule.	(1) CCOC (2) NDP
2.02	1/01.08.2019	T	two weeks schedule check	CCOC should hand over the two weeks plan.	CCOC
2.03	1/01.08.2019	T	main materials and equipment	(1) Prepare the materials for approval (2) List of all vendors.	NDP

Items	Meeting No./Date	Importance(T/I)	Delays		Responsible
2.01	2	T	excavation (delayed)	The excavation should begin before...	CCOC 21.01.2019
2.02	2	T	ground survey (delayed)	The geological survey report should be delivered before...	CCOC 25.01.2019
3.01	2	T	the temporary office (delayed)	Building hasn't been finished yet.	(1) CCOC (2) NDP

（趁热打铁，课后练习见150页）

Answers

Chapter A Weekly Meetings

Ⅱ Read the Material Below and Answer the Questions

1. On the 18th of January and the 25th of January 2019.
2. The management company needs to check the list we sent them first.

Chapter B Delayed

Ⅱ Read the Material Below and Answer the Questions

1. The speakers are talking about the delayed works shown on the meeting minutes.

2. The delays are: (1)The excavation hasn't been started yet. Because the approval hasn't been approved yet. (2)The geological report hasn't been handed in. Because the situation is far more complicated than it was expected. (3)The temporary office hasn't been completed because of the late delivery of materials.

Chapter C Explaining

Ⅱ Read the Material Below and Answer the Questions

The entire schedule of the project would not be delay, because CCOC has taken measures to cope with the problems at hand.

Ⅳ Exercise and Activity

假设在项目上因遇到以下问题造成延迟,你会怎么做?(仅供参考)

The workers have been working overtime on site for a week! Still the schedule is behind!	Ask for a favor to extend the deadline.
The delivery is late, and the building hasn't been started yet!	Give more pressure to the vendor, and sign a contract to secure the time.
The owner wants to change our project manager.	Discuss this matter in a meeting first.
Waiting for government approval.	Hope your owner understands the process of the government.
Quality issues with our goods.	Assure them with test results, and change the brand if possible.

After-class Exercises
课后练习

Class 1 After-class Exercises

1. 请翻译以下句子。

我们将要在103地区建设一个全新的科技研究中心。(science and research center)
爸爸我正忙着写设计方案呢,我今晚下班给你回电话。
CNPC公司是一个非常成熟的工程公司,我们计划在3年后成立6家子公司。(sub-branches)
Johnson正在开着会,Mr. King 正在面试一名新员工,我建议你下午再来找他们比较好。

2. 请用左列提出的计划造句。

clarification meeting tomorrow morning	
dinner tonight at 6:30 p.m.	
pick up our client at 10:00 p.m. in the airport	
hand in the samples (样品) before July	

3. 请根据要求回答。

Can you briefly introduce yourself to us?(use present, continuous, and future tense)

4. 请根据要求回答。

A foreign client is coming to your company for a visit; ask him some open questions to know more about him. (work/hometown/education/free time)

Class 2 After-class Exercises

1. What did you do at work or during the given time below?

Last Saturday	
Yesterday	
Monday last week	
Year 2016	
December 2017	

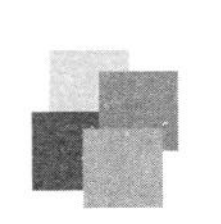

2. 请用完成时来回答以下问题。

Have you achieved anything great in your career?	
What projects have you done?	
Have you talked to any foreign clients before?	
Have you been to anywhere special lately?	
Why have you chosen your job?	

3. 请用过去时和完成时介绍以下机构的背景。

your current company or school

the biggest organization in your field

Class 3 After-class Exercises

1. 请写出下列词语的反义词。

honest (dishonest)	respectful	mature	important
professional	reliable	polite	confident
popular	influential	responsible	cheap
kind	optimistic	organized	trustworthy
wise	experienced	decisive	rude

2. 请将下列句子变为从句。

CNNC is a really mature company. CNNC has 10 years of experiences in the engineering field.
Our service includes EPC. It covers fields from chemical industry to pharmaceutical industry.
I use this product every day. It's really famous in our country.

3. 用适当的形容词来描述左列中的人物。

a project manager	Should be really reliable and professional.
an EHS engineer	
a general manager	
a sales manager	
a designer	
the security staff	

4. 将下列表述转化为看法。

We waste too much water in the project.	
Jason doesn't want to work here any longer.	
This man need a rest; he has been working for 20 hours!	
Our company is better than that company!	

Class 4 After-class Exercises

你会如何介绍以下图片?

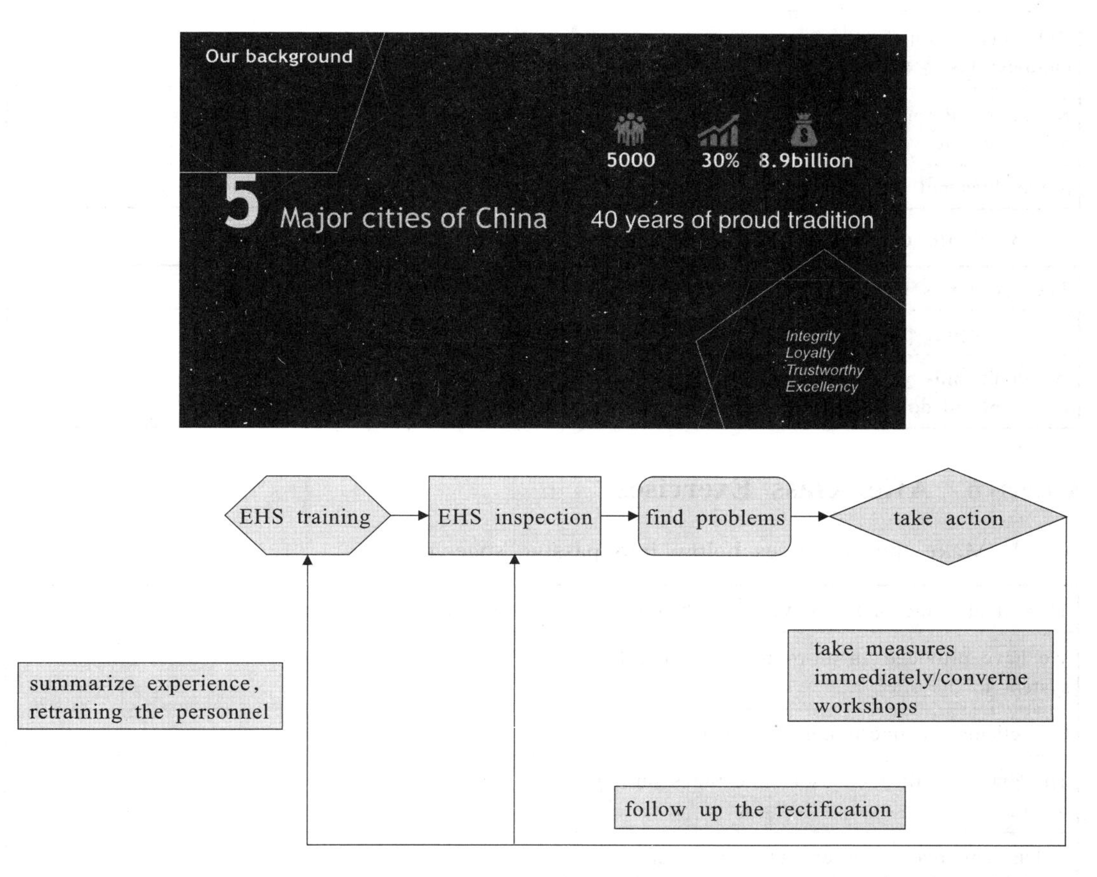

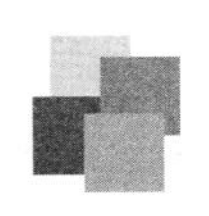

Class 5 After-class Exercises

1. Provide replies for the email below.

To Whom it may concern,
I'm writing this email to inform you that one of our power cord has power outage, and it has already caused some power shut down in our offices, and also, please check the pipes upstairs, which seem to be leaking a lot of water out. Please tackle this issue as soon as possible before anything happens.
Sincerely yours

2. Solve these problems below.

We cannot pay the investment on time.	How come? What happened? It seems that this might violate our contract!
Bob was arguing with our engineer last week.	
Kevin is not satisfied with the price we offer.	
Steve does not like his job.	

3. Write down the "orders" towards the problems below.

The rooftop is leaking.	The rooftop needs to be fixed.
The firefighting pipes are broken.	
We don't think your company is capable of doing this project.	

Class 6 After-class Exercises

1. Make the sentences below into passive speech.

They held a meeting last week in Shanghai.	A meeting...
We have provided an all-round pest control system to the site.	
Our clients are questioning our quality.	
The owner challenged us couple times during the bidding.	
Helen took over Jennifer's work yesterday.	

In order to maintain the standard of our quality and safety, our company has made great efforts on the quality trainings, and we held a seminar today morning to enhance our safety procedures, while we also consider energy saving an important part of our work.

3. Make up a short report about the picture below.

Class 7 After-class Exercises

1. What would you do if...

If you found a really serious problem on the site?
I would...
If you were the boss of your company?
If your colleagues damaged a really expensive piece of equipment?
If you won a really huge bid?

2. What should they have done instead?

Saturday last week, our manager team went to Shanghai for an important commercial discussion with KSC, and our client refused to discuss any details, but just asked us to give out our final price towards the project, and we can not make any changes thereafter. Our teammates weren't ready for it and gave him a really low price. In the end, although we won the project, we cannot operate properly due to the insufficient investment.

3. Speculate the things that might have happened below.

Your owner didn't call you after being late for 2 hours.
You waved (招手) to your most important client in a party and said hello, but he kind of ignored you.
One of your new recruits has been absent for 3 weeks without any reasons.

Class 8 After-class Exercises

1. Write down your replies to the questions below.

So, what do you want to eat tonight? Or you want me to pick a restaurant for you?
Ummm, the food is really delicious here; do you come here a lot?
I want something made from chicken and beef, got any recommendation?(推荐)

2. 请用am/is/are supposed to 或者 was/were supposed to (was/were going to)来翻译以下句子。

我们今晚应该会有一场晚宴。	We are supposed to...
昨天我们本来应该和客户见面,但是他没来。	
明天我们应该要开一场安全会议。	
我们本来应该在月初就能完成这个项目了。	
今天应该会下雨,我们要做好准备。	
后天本来我们要去美国开会,但是业主改了主意。	

3. 你的好友Patrick刚刚升职,请向他发送一封祝贺信。

Dear Patrick,

Class 9 After-class Exercises

1. 描述下方图表。

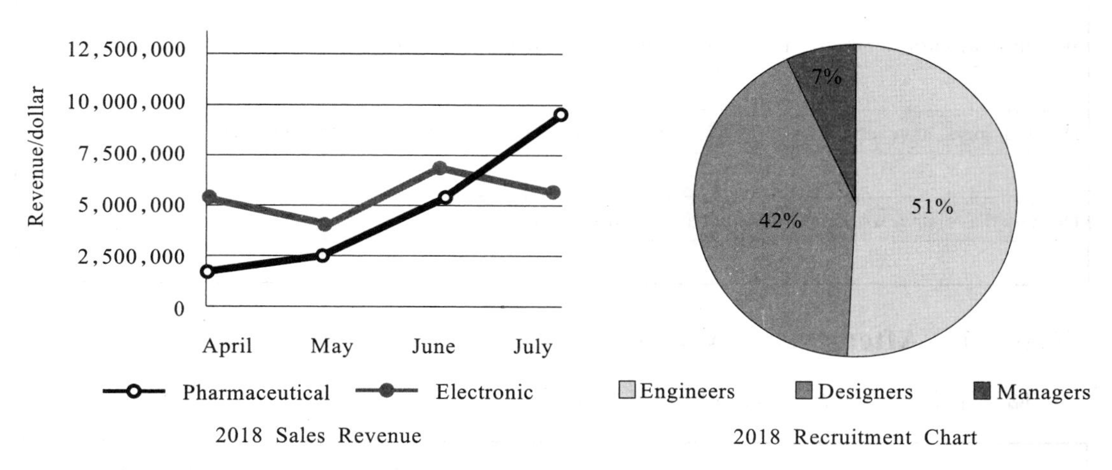

2018 Sales Revenue

2018 Recruitment Chart

2. 请将下列形容词变为有比较属性的形容词。(可用more/less)

large (larger)	professional	early	fast
interesting	reliable	safe	clean
competitive	good	powerful	experienced
long	expensive	risky	dangerous

Class 10 After-class Exercises

1. Find a partner and interview him/her. Make questions and record the answers. (Try to answer them using the same tenses, and try to use more than one tense to answer these questions.)

How long have you been working in this field?	I have been...
When did you ______?	
Have you ______?	
What is your favorite ______?	
Do you still remember when ______?	
How many years have you been ______ in this company?	
Why did you choose ______?	
What are you going to ______?	
What would you do, if ______?	
Do you regret on anything you have done?	
What are you busy doing lately?	
What do you think about yourself and your department (or your job)?	

2. Make the sentences below into reported speech.

From today on, every engineer on the site must have an authorization card to get in!
We must discuss about the recent problem towards the quality on the installation.
Do you guys have any clue (是否知晓) when we are going to finish this project?
Do you have any workers I can borrow to my project?

Class 11 After-class Exercises

阅读下方会议纪要,完成下周开幕式计划。

1. It's very important that the ground breaking ceremony should be prepared before 3:00 p.m., on August 21st 2018.
2. KDP will pick up the government officials from the airport at 9:00 a.m. on the 21st of August.
3. Gathering all the workers to the ceremony at 2:30 p.m. on the 21st of August.
4. Check the food and Champaign.
5. Get the vehicles starting (check).
6. The NDP company will do an all-round inspection on safety before 3:00 p.m.
7. The ceremony begins.

Item	Meeting No./Date	Importance(T/I)	GC Contract Issues	Responsible
1.01				
1.02				
1.03				
1.04				
1.05				
1.06				
1.07				

Answers for the After-class Exercises

课后练习答案

Class 1 After-class Exercises

1. 请翻译以下句子。

我们将要在103地区建设一个全新的科技研究中心。(science and research center)
We are going to build a brand new science and research center in district 103.
爸爸我正忙着写设计方案呢,我今晚下班给你打回去。
I'm working on my new designs dad; I will call you after work.
CNPC公司是一个非常成熟的工程公司,我们计划在3年后成立6家子公司。(sub-branches)
CNPC is a really mature company, and we are going to establish 6 more sub-branches within the next three years.
Johnson正在开着会,Mr. King 正在面试一名新员工,我建议你下午再来找他们比较好。
Mr. Johnson is having a meeting right now, and Mr. King is currently interviewing a new recruit. So, I suggest you look for them in the afternoon.

2. 请用左列提出的计划造句。

clarification meeting tomorrow morning	We are going to have a clarification meeting tomorrow afternoon.
dinner tonight at 6:30 p.m.	Our workmates are going to have a dinner at 6:30 p.m. today.
pick up our client at 10:00 p.m. in the airport	Paul is going to pick up a client at 10:00 p.m. in the airport.
hand in the samples (样品) before July	Our company is going to hand in the samples before July this year.

3. 请根据要求回答。

Can you briefly introduce yourself for us?(use present, continuous, and future tense)
Hey guys, I'm Dawson Lee. I think I'm pretty outgoing and practical. I have been working in CCOC for the past 13 years and have done projects such as KDP and CNPN. Nice to meet you guys.

4. 请根据要求回答。

A foreign client is coming to your company for a visit; ask him some open questions to know more about him. (work/hometown/education/free time)
How is your work recently? So, what do you think about our company? Can you tell us more about your demands?

Class 2 After-class Exercises

1. What did you do at work or during the given time below?

Last Saturday	I was having a drink in the city central last Saturday.
Yesterday	We had a team building yesterday, and we enjoyed it very much.
Monday last week	Our department organized a soccer match on Monday last week.
Year 2016	On the year 2016, we established our first global engineering center.
December 2017	This new recruit graduated from the college in December 2017.

2. 请用完成时来回答以下问题。

Have you achieved anything great in your career?	I have been studying in the US for the past 7 years, and have obtained 3 degrees and 2 civil engineering awards.
What projects have you done?	Our team has done many international projects such as the KDFM and CPS, and our services are the best in the industry.
Have you talked to any foreign clients before?	I can't really remember when, but we definitely have talked to the owner of KDM.
Have you been to anywhere special lately?	I have been to Italy for vacation, and it was cool.
Why have you chosen your job?	I have had an appointment with my boss, and he has convinced me to come.

3. 请用过去时和完成时介绍以下机构的背景。

your current company or school
CCOC is a practical engineering company that has been standing in the engineering industry for the past 18 years...

the biggest organization in your field
NDP is the biggest engineering company in China, and their goal is to become a high-end international construction company.

Class 3 After-class Exercises

1. 请写出下列词语的反义词。

honest (dishonest)	respectful (disrespectful)	mature (immature)	important (unimportant)
professional (unprofessional)	reliable (unreliable)	polite (impolite)	confident (shy)
popular (unpopular)	influential (uninfluential)	responsible (ilresponsible)	cheap (expensive)
kind (unkind)	optimistic (pessimistic)	organized (disorganized)	trustworthy (untrustworthy)
wise (ignorant)	experienced (inexperienced)	decisive (indecisive)	rude (polite)

2. 请将下列句子变为从句。

CNNC is a really mature company. CNNC has 10 years of experiences in the engineering field.
CNNC is a really mature company that has 10 years of experiences in the engineering field.
Our service includes EPC. It covers fields from chemical industry to pharmaceutical industry.
Our service includes EPC, which covers fields from chemical industry to pharmaceutical industry.
I use this product every day. It's really famous in our country.
I use this product every day that is really famous in our country.

3. 用适当的形容词来描述左列中的人物。

a project manager	Should be really reliable and professional.
an EHS engineer	Should be really careful and precise.
a general manager	Should be smart and organized.
a sales manager	Should be outgoing and practical.
a designer	Should be versatile and talented.
the security staff	Should be smart and hardworking.

4. 将下列表述转化为看法。

We waste too much water in the project.	It seems to me...
Jason doesn't want to work here any longer.	If you ask me...
These men need a rest; he has been working for 20 hours!	I think...
Our company is better than that company!	I don't really think that...

Class 4　After-class Exercises

你会如何介绍以下图片？

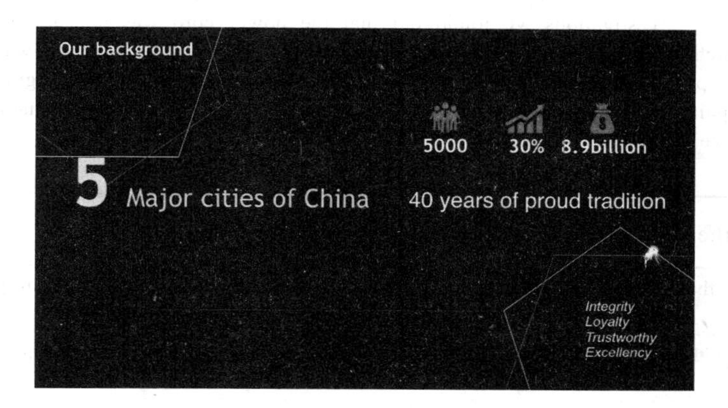

So this page is about our company's background. We have 5,000 employees and a guaranteed 30% growth each year, and last year we have reached 8.9 billion RMB worth of contract value. If you ask me, CCOC is a strong and mature corporation.

We have been dedicating in the engineering industry for the past 40 years with the proud tradition of integrity, loyalty, being trustworthy and excellency, and we are going to deliver the most outstanding services for every each one of our clients.

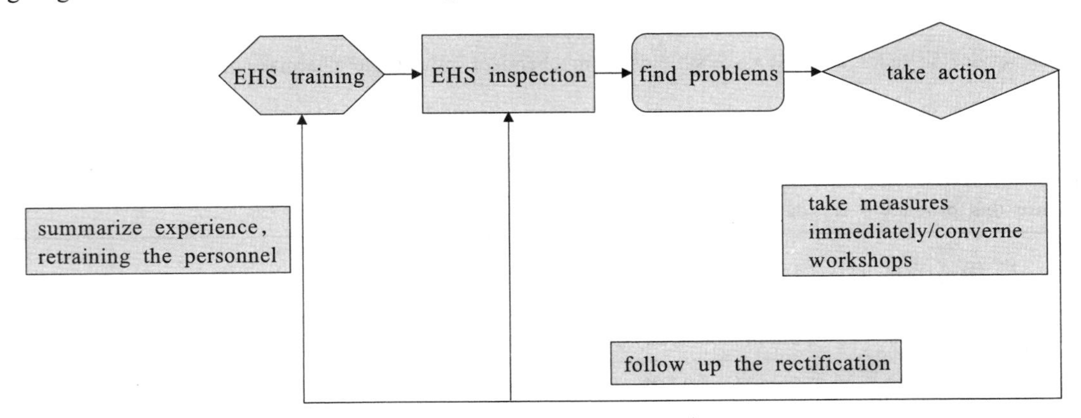

This page is our company's EHS procedure, from EHS training to EHS inspection. And if we find problems, immediate actions will be taken as well as measures, then follow up the rectifications and summarize our experience for future trainings.

Class 5 After-class Exercises

1. Provide replies for the email below.

To Whom it may concern,
I'm writing this email to inform you that one of our power cord has power outage, and it has already caused some power shut down in our offices, and also, please check the pipes upstairs, which seem to be leaking a lot of water out. Please tackle this issue as soon as possible before anything happens.
Sincerely yours

Dear sir/madam,
We are writing this letter to apologize for any inconvenience that happened in your offices. However, after our investigations we found out that the power cord was not properly assembled by your staffs, which is the main cause of the outage, also the water leakage was caused by an outdated broken pipe. It's unfortunate that this has happened, but we are going to send our engineers to repair it within 24 hours free of charge, would it be OK? Thank you for your concern, and have a wonderful day.
Best regards

2. Solve these problems below.

We cannot pay the investment on time.	How come? What happened? It seems that this might violate our contract!
Bob was arguing with our engineer last week.	What happened? There must be a misunderstanding! I think Bob is a quiet guy who doesn't like to talk!
Kevin is not satisfied with the price we offer.	Well it seems to us that these brands we recommended are of the highest quality. It's a good idea to make some comparison to other brands.
Steve does not like his job.	How come? I thought that he just had a promotion.

3. Write down the "orders" towards the problems below.

The rooftop is leaking.	The rooftop needs to be fixed.
The firefighting pipes are broken.	We need to get Mr. Wall to handle this problem.
We don't think your company is capable of doing this project.	It seems to me that discussions must be done before quick decisions.

Class 6 After-class Exercises

1. Make the sentences below into passive speech.

They held a meeting last week in Shanghai.	A meeting was held last week in Shanghai.
We have provided an all-round pest control system to the site.	An all-round pest control system has been provided.
Our clients are questioning our quality.	Our project quality was being questioned by our clients.
The owner challenged us couple times during the bidding.	We have been challenged a couple times by the owner during the bidding.
Helen took over Jennifer's work yesterday.	Jennifer's work was taken over by Helen yesterday.

2. Make the following dialogue into passive speech.

In order to maintained the standard of our quality and safety, our company has made great efforts on the quality trainings, and we held a seminar today morning to enhance our safety procedures, while we also consider energy saving an important part of our work.
Great efforts have been made by our company to maintain the standard of our quality, and a seminar was held today to enhance the safety procedures, while energy saving has also been considered an important part of our work.

3. Make up a short report about the picture below.

A safety meeting was held yesterday in the office. Engineers were asked to report their project safety status, and they were asked to provide improvement suggestions towards the project.

Class 7 After-class Exercises

1. What would you do if...

If you found a really serious problem on the site?
I would report this immediately to our head office, and make some first aid actions as soon as possible.
If you were the boss of your company?
I would certainly increase the portion of our international clients.
If your colleagues damaged a really expensive piece of equipment?
I would probably call our boss to find some solutions.
If you won a really huge bid?
We would totally celebrate for it.

2. What should they have done instead?

Saturday last week, our manager team went to Shanghai for an important commercial discussion with KSC, and our client refused to discuss any details, but just asked us to give out our final price towards the project, and we can not make any changes thereafter. Our teammates weren't ready for it and gave him a really low price. In the end, although we won the project, we cannot operate properly due to the insufficient investment.

Well, I think we should have prepared earlier before we meet up with our clients. We should have had some inspections towards the project and should have discussed and re-calculated(重新计算) our budget with our general manager. Anyhow, we are going to cope with this problem.

3. Speculate the things that might have happened below.

Your owner didn't call you after being late for 2 hours.
He might have been really busy and could have had a meeting with someone important.
You waved to your most important client in a party and said hello, but he kind of ignored you.
He must have been thinking about something else, and didn't notice me.
One of your new recruits has been absent for 3 weeks without any reasons.
He might have found a new job or something bad could have happened to him.

Class 8 After-class Exercises

1. Write down your replies to the questions below.

So, what do you want to eat tonight? Or you want me to pick a restaurant for you?
Emm, I'm supposed to complete my documents before 6:00 p.m., so I think you can book it at around 6:30.
Ummm, the food is really delicious here; do you come here a lot?
Well, this is the first time for me too; my best friend recommended it to me.
I want something made from chicken and beef, got any recommendation?
Sir, we have a really delicious curry chicken that I think you can give it a try.

2. 请用am/is/are supposed to 或者 was/were supposed to (was/were going to)来翻译以下句子。

我们今晚应该会有一场晚宴。	We are supposed to have a banquet tonight.
昨天我们本来应该和客户见面,但是他没来。	We were supposed to meet our client yesterday, but he didn't come.
明天我们应该要开一场安全会议。	There is supposed to be an EHS meeting tomorrow.
我们本来应该在月初就能完成这个项目了。	The project is supposed to have been finished earlier this month.
明天应该会下雨,我们要做好准备。	We are supposed to get ready for the rain tomorrow.
本来我们要去美国开会,但是业主改了主意。	We were going to have a meeting in the US, but our owners changed their minds.

3. 你的好友Patrick刚刚升职,请向他发送一封祝贺信。

Dear Patrick,
I'm writing this letter to congratulate on your promotion. These three years, I have seen all the hard works you have done, and all the sacrifices you have made for our company, and I think you totally deserve this promotion. Again, congratulations on your promotion. Wish you all the best in your future career. Also, we are throwing you a party this Saturday evening, please come, and don't forget to bring your wife and kids!
Best Regards
Richard Lewis

Class 9 After-class Exercises

1. 描述下方图表。

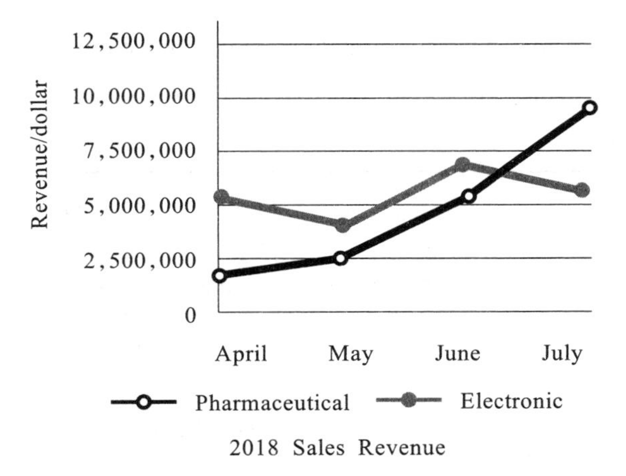

2018 Sales Revenue

So this is the graph for the 2018's sales revenue from April to July for our company. As you can see on the chart, the revenue for the pharmaceutical department has been growing rapidly, and it reached 10 million dollars by the end of July.

However, as for the revenue of the electronic department there has been ups and downs during the last four months, and we are going to find out the problems and hopefully turn it up as soon as possible.

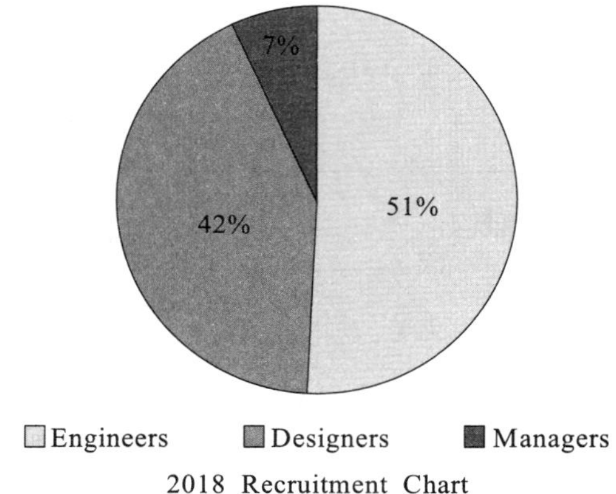

2018 Recruitment Chart

Please take a look at the chart above. This chart shows our company's portion of human resource. Our engineers have the biggest portion of 51%, following the designers 42%, and as for the managers, the total portion is about 7%.

2. 请将下列形容词变为有比较属性的形容词。(可用 more/less)

large (larger)	more professional	earlier	faster
more interesting	more reliable	safer	cleaner
more competitive	better	more powerful	more experienced
longer	more expensive	riskier	more dangerous

Class 10 After-class Exercises

1. Find a partner and interview him/her. Make questions and record the answers. (Try to answer them using the same tenses, and try to use more than one tense to answer these questions.)

How long have you been working in this field?	I have been working in this field for the past 17 years.
When did you __________?	When did you graduate from the college?
Have you __________?	Have you worked in any English speaking country?
What is your favorite __________?	My favorite job is managing my EHS teammates.
Do you still remember when __________?	Actually I do. I was really nervous during my first interview.
How many years have you been __________ in this company?	I have been directing the production of this company since 1994.

Continued

Why did you choose ____________?	I chose this work because my major in college suits it.
What are you going to ____________?	I'm going to have a thorough discussion with the owner about the vendor's list at 3 p.m.
What would you do, if ____________?	I would ask for our company's support, if the project were delayed.
Do you regret on anything you have done?	I should have been to the college when I had a chance.
What are you busy doing lately?	Well, I'm trying to complete my annual task lately, and it's not going so well.
What do you think about yourself and your department (or your job)?	It seems to me that my department has great potential in the industry, and we are working really hard to make more progress for it.

2. Make the sentences below into reported speech.

From today on, every engineer on the site must have an authorization card to get in!
Our boss said every engineer on the site must have an authorization card to get in from that day on.
We must discuss about the recent problem towards the quality on the installation.
Our owner said we must discuss about the recent problem towards the quality on the installation.
Do you guys have any clue when we are going to finish this project?
Mr. Mathew asked if we had any clue when we were going to finish this project.
Do you have any workers I can borrow to my project?
The project manager asked if I had any workers he could borrow to his project.

Class 11 After-class Exercises

阅读下方会议纪要,完成下周开幕式计划。

1. It's very important that the ground breaking ceremony should be prepared before 3:00 p.m., on August 21st 2018.
2. KDP will pick up the government officials from the airport at 9:00 a.m. on the 21st of August.
3. Gathering all the workers to the ceremony at 2:30 p.m. on the 21st of August.
4. Check the food and champagne.
5. Get the vehicles starting (check).
6. The CCOC will do an all-round inspection on safety before 3:00 p.m.
7. The ceremony begins.

Item	Meeting No./Date	Importance (T/I)	GC Contract Issues	Responsible
1.01	5th of August,2018	I	The information of the ceremony	CCOC
1.02	5th of August,2018	T	Pick up the government officials from the airport at 9:00 a.m.	KDP
1.03	5th of August,2018	T	Gathering all the workers to the ceremony at 2:30 p.m.	CCOC
1.04	5th of August,2018	T	The CCOC will do an all-round inspection on safety before 3:00 p.m. (the ceremony starts)	CCOC
1.05	5th of August,2018	I	Check the food and champagne	CCOC
1.06	5th of August,2018	I	Get the vehicles starting (check)	CCOC